BORDERLAND

BORDERLAND

WALKING THE EAST KENT COASTLINE

ROD EDMOND

Matador
Unit E2 Airfield Business Park
Harrison Road, Market Harborough
Leicestershire LE16 7UL
Tel: 0116 279 2299
Email: books@troubador.co.uk
Web: www.troubador.co.uk/matador
Twitter: @matadorbooks

ISBN 9781803136783

British Library Cataloguing in Publication Data.
A catalogue record for this book is available from the British Library.

Printed and bound by CPI Group (UK) Ltd, Croydon, CR0 4YY
Typeset in 11pt Adobe Garamond Pro by Troubador Publishing Ltd, Leicester, UK

Matador is an imprint of Troubador Publishing Ltd

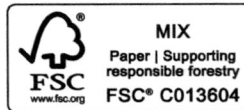

For Abbas, Abdelatif and Ranjit

'The coast of Kent is England's chin,
Jutting out to take the punches'
(Anon)

'And if a stranger sojourn with thee in your land, ye shall not vex
him… the stranger that dwelleth with you shall be unto you as one
born among you'
(Leviticus)

'I love crossing borders'
(Olga Tokarczuk, *Drive Your Plow Over the Bones of the Dead*)

CONTENTS

‡

ILLUSTRATIONS

‡

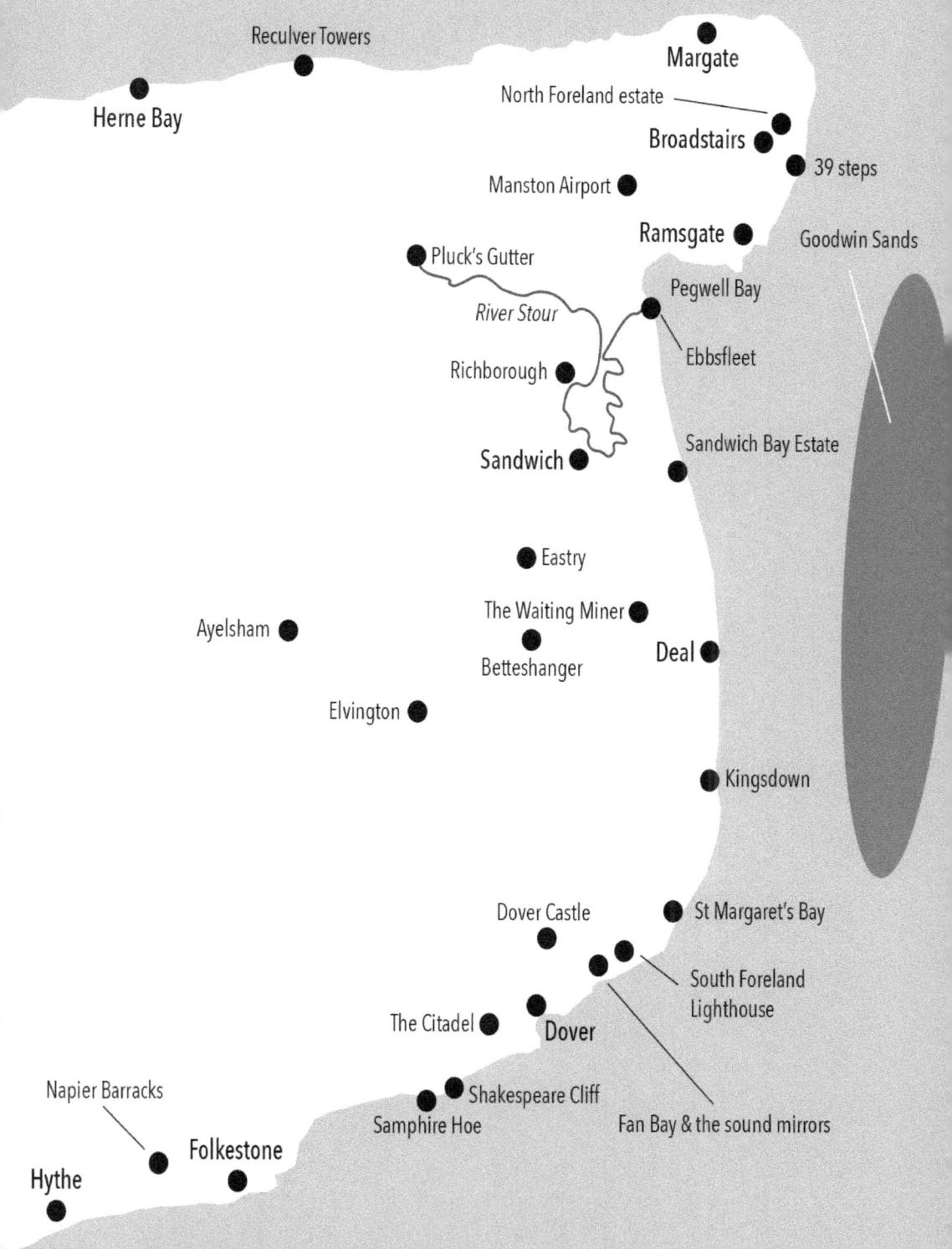

Reculver Towers

Margate

Herne Bay

North Foreland estate

Broadstairs

39 steps

Manston Airport

Ramsgate

Goodwin Sands

Pluck's Gutter

Pegwell Bay

River Stour

Ebbsfleet

Richborough

Sandwich Bay Estate

Sandwich

Eastry

The Waiting Miner

Ayelsham

Deal

Betteshanger

Elvington

Kingsdown

St Margaret's Bay

Dover Castle

South Foreland
Lighthouse

The Citadel

Dover

Napier Barracks

Shakespeare Cliff

Samphire Hoe

Fan Bay & the sound mirrors

Hythe

Folkestone

PRELUDE

‡

I knew very little about the Goodwin Sands until I almost drowned there. Although I'd lived in Canterbury for more than thirty years, I was even vague about where they lay. Somewhere off the coast of Thanet was all I knew. So, when the BBC series *Coast* enlisted my cricket team, Beltinge, to re-enact an old tradition of playing a match on the Sands, I thought, *Why not?*

I did some hasty internet research. I'd forgotten that they figured in *The Merchant of Venice* as that 'very dangerous flat and fatal, where the carcases of many a tall ship lie buried', where one of Antonio's richly laden ships had run aground. I discovered that the earliest record of cricket on the Sands was a watercolour by Turner, and I found an *Illustrated London News* report and woodcut of a match in 1854 between 'Captain Pearson and 10 members from his lugger, the *Spartan*' and 'a party of Walmer Gentlemen'. There were reports of later matches too, one from 1919 in which the participants narrowly escaped drowning, another in 1973 between the ratings and officers of the Royal Navy Inshore Fleet. I took little heed of the near drowning; 1919, after all, was a long time ago. I liked the connection with Turner, and the grainy woodcut in the *Illustrated London News* depicted a calm and benign setting with picnickers on the Sands and the mainland comfortingly nearby. I also liked the flavour of that long tradition in English cricket of gentlemen versus players in the line-up of these encounters – the crew of a lugger or naval ratings on one side, Walmer gentlemen or naval officers on the other.

Beltinge eschewed such distinctions. We were divided by age rather than class, a mix of promising young cricketers soon to be lost to the club as they escaped the Isle of Thanet and ageing ones who were no longer quite the players they continued to think they were. Beltinge often started its Sunday matches a couple of players short as latecomers drifted in after a heavy Saturday night or arrived hotfoot from delivering young children to grandparents or a birthday party. On this particular morning though, everyone had turned up outside the maritime museum in Ramsgate harbour sharp at 6.30am, having been warned that an early start was necessary if we were to catch low tide on the Sands. Unusually too, everyone had their full kit, the prospect of being on TV resulting in a more professional-looking outfit than normally graced the East Kent Village League. The BBC crew, on the other hand, turned out as if they were at Beltinge Cricket Club on a Sunday afternoon. It was gone 7am before the camera team had finished its breakfast and straggled down the hill from their hotel to the harbour. After further messing about, we finally set off for the Goodwin Sands more than an hour late. But it was a beautiful morning and the leisurely embarkation felt somehow in keeping with the spirit of the re-enactment. I also assumed the BBC knew what it was doing.

Our journey took longer than I'd expected but the water was calm; the sun was shining; and all was well with the world. After half an hour or so the Sands loomed up rather like a whale just as it is starting to breach. We anchored fifty yards or so off the Sands and transferred into a RHIB, a rigid-hulled inflatable boat, that took us into the shallows from where we waded ashore. You can see us there at the end of the second series of *Coast*, emerging from the water in cricket whites, with bats and pads under our arms, like less scantily clad Honey Ryders gazing with wonder at the strange new world of the Goodwins. It was warm and very bright, the sun reflecting off the pools of seawater that pitted the surface of the Sands. In the distance I could make out the shapes of basking seals.

Naively I had imagined a robust game of beach cricket such as I'd enjoyed as a child in my native New Zealand but instead, we were extras in a film set. A single set of stumps was placed on a narrow strip of firm, gently sloping sand near the water's edge. We took up fielding positions; one of the team pretended to bowl; the programme's presenter, Neil Oliver, played an air shot to leg and I took a fake catch, the ball lobbed to me from behind the crouching figure of the cameraman. With a haka-like appeal from the slips that threatened to put several of our team out of the coming Sunday's match, the shoot was done. Virtual cricket had reached the Goodwins.

What followed however was real enough. It is often remarked that cricket is an unlikely game to have been born in a country with a climate like England's. A broken afternoon with short spells of play squeezed in between showers, or a run-chase against the weather with thirty runs still needed and the drizzle thickening to rain, is familiar to all cricketers. Cricket on the Goodwin Sands was an extreme version of this tension between the sport and the weather. Play had started an hour late and we had only a short time before the tide would rush back.

Suddenly there were shouts from the RHIB. The tide had turned, and we must get back to the boat immediately. We hurriedly splashed our way through the rising waters while the BBC crew, urgent for the first time that morning, gathered up its equipment. The speed of the incoming tide was extraordinary, like a wave after it has broken and surged ashore, but without the reflux that follows. It came on relentlessly and from all sides so that the currents around the boat were soon colliding and recoiling. We clambered into the RHIB but the massed weight of the cricket and television teams, together with the cricket kit and the filming equipment which had been taken onto the Sands in a separate landing, meant that we stuck side-on to the incoming tide. Some of us got back into the water, which was now chest-deep, to try and swing the boat around to face the main current of the tide, but it was caught fast in the Sands.

The skipper of the boat decided we should sit tight until floated off by the tide. But the weather had changed as the tide rushed in and a strengthening north-easterly was whipping up waves that broke over the side of the boat, filling it with water and sinking it deeper in the Sands. With the water in the boat over our waists, waves dumping on us every few seconds, and the flotsam of bats, pads, mics and other equipment starting to be washed away, one of our team had a panic attack. Until now I'd been discomfited rather than frightened, but as the Goodwins sank from view, the horizon lengthened and the moans of my teammate rose, I became alarmed. Even our garrulous cameraman, who had earlier spoken casually of having been under fire in Northern Ireland, fell quiet.

For some reason the skipper was reluctant to put out a mayday call. Perhaps he found the idea of being rescued galling but there was no such pride at stake for the rest of us. He revved and revved the outboard motor to try and break free of the Sands, but this merely increased the turbulence around the boat and embedded us more firmly. The harsh repetitive noise of the revving added a nerve-grating soundtrack to the scene in which we were caught. I was by now very cold as well. Eventually, much later than reason suggested, the driver put out a call for help. Our wait for the RNLI, the Royal National Lifeboat Institution, was probably shorter than it felt. A plane circling overhead was reassuring insofar as it showed we were known to be in danger, but it was tantalisingly unable to help. Eventually, a lifeboat arrived from Dover; a line was attached to our craft; and we were pulled off the Sands and transferred to a larger vessel. There's a brief, rather grainy shot of us climbing up the side of this vessel at the end of the *Coast* episode, the voice-over blandly reflecting on how things can go wrong and sometimes do. Re-entering Ramsgate harbour had the feel of returning to the pavilion after an ignominious duck, although even on English cricket grounds in April I had never felt so cold. I drove home to Canterbury, filled the bath with hot water and lay there thawing. That Sunday Beltinge lost heavily.

The BBC lost several hundred thousand pounds of equipment but managed to save the film. Beltinge lost most of its gear but the BBC, embarrassed by the danger into which it had dropped us, and no doubt worried about legal action because of its failure to make a reconnaissance of the Sands and its negligence in our delayed start, made sure that we were immediately compensated with new kit. The Goodwin Sands had added television and cricket gear to the wreckage of centuries held in its depths. W.H. Auden, in his poem 'In Sickness and Health', compares the way in which love is prone to delusion with 'set(ting) up shop on Goodwin Sands'. Our ill-prepared re-enactment had been marked by a similar credulity.

It was not just the Goodwin Sands that I'd ignored while living for many years in Canterbury but the whole East Kent coastline. As far as I was concerned, Kent's littoral was another world. Dover was just a place of transit, Sandwich a town where the British Open was held every few years, and Thanet, its most prominent headland, was, as it is often called, 'Planet Thanet'. In Canterbury I lived facing London with my back to the coast. Soon after my rescue from the Goodwin Sands, however, I began to move coastwards, first to Ash, a village just inland from Sandwich, and then to Deal on the edge of the English Channel a few miles north of Dover. This was prompted by a number of changes in my life. I retired, a little early, from my academic post at the University of Kent. My relationship of more than thirty years ended. My father in New Zealand died. Suddenly I was cut adrift from the moorings of my life which had held me in place for decades. I embarked happily, a little anxiously, on a new relationship and set out in my mid-sixties to make a new life, though one heavy with the old.

Moving house involves more than bricks and mortar and the shedding of what one no longer wants, or needs, or can face. Psychologically and ritually, it touches much deeper things, life and death stuff. I reread Richard Mabey's *Nature Cure* (2005) which described how moving from his home in the Chilterns, where he'd always lived, to the flatlands of East Anglia had rescued him from depression and opened up a new existence. This became, at first, a model of sorts, although there were many differences in our situation. I wasn't ill, just tired and stale. Susan Sontag wrote that while 'some people are their lives', others 'merely inhabit them'. I felt that I'd been renting mine and it was now time that I owned it.

Mabey moved to a setting he'd known as a child so for him it was like a return to his beginnings and a better time. My only close experience of the East Kent coastline, near-death, was less promising. And although I'd lived in England for most of my adult life, I'd remained a resident alien, a metic. I'd come to England in the late 1960s for postgraduate study and stayed, but in my bones, I remained a New Zealander. I'd never taken a British passport and my only secure foothold in the United Kingdom was an insignificant-looking stamp – 'Given leave to enter the United Kingdom for an indefinite period' – in an expired passport that I must carry with me whenever I left and re-entered the country. After more than four decades of residence, my tenure in England still felt uncertain. Indeed, since the Immigration Act 2014, my stamp no longer automatically protected me from deportation.

I didn't much like the East Kent coast at first. Arriving at any coast, whether by sea or land, usually quickened my spirit, 'renewed my blood' as Robert Louis Stevenson puts it at the opening of his story *The Beach of Falesá*, but this shore seemed flat and drab and featureless, apart that is from the cliché of the White Cliffs. I love sand, but the coastline from Sandwich to Dover was pebbled, difficult to walk over and uncomfortable to lie on. I'm never more homesick for New Zealand than when by the sea, remembering the long stretches of white sand, rocky bays lined with pohutukawa trees, rolling surf, golden days and barefoot summers I took for granted as a child. As George Orwell said, 'In one's childhood it never rained'. The only parts of the British coastline I'd ever much liked were where my nineteenth-century forebears had come from: Cornwall, the West Highlands of Scotland, the North Sea coast of Aberdeenshire. These places felt somehow familiar, as if they were in my DNA. I had no instinctive sympathy, no kinship with the exposed and windswept littoral where I'd come to live.

In feeling this, I was doubtless influenced by W.G. Sebald's seductive

The Rings of Saturn (1998) whose much-quoted phrase, 'the east stands for lost causes', sounds the keynote of how the coast of East Anglia is now often regarded. The shoreline Sebald walks in that book is eerily without people, as if its inhabitants have been lost in the genius of the place, bested by its integral character. At first, I saw the Kent coastline as an extension of the East Anglian, MAMBA country: Miles and Miles of Bugger All. And the White Cliffs were tainted by the insular idea of nationhood they are so often enlisted to express.

With time, however, this instinctive resistance eased and the land- and seascapes of my new location began to filter through my senses into my being. How does location, mere place, become home ground? Curious to know better where I was, and hoping this would make me feel more at home in the place that was now my home – empty land under empty skies as it often seemed – I went walking. It's difficult to get lost while walking a coastline. You just keep the sea to one side of you. There's no poring over maps or way signs. You can look up and around while continuing to walk vigorously. I find slow walking frustrating. A line from the mid-twentieth-century poet and doyen of New Zealand letters, Charles Brasch, kept coming back to me: 'I tramp my streets into recognition'. Brasch had an uneasy, Anglo-centric relation with his home culture, and this line, from a late poem 'Home Ground', catches the strain and effort in making his country live in him. Kent, England wasn't even my country, and it was shoreline and clifftops I walked, not streets, but like Brasch, I tramped and sweated to occupy this place so that it became familiar, even perhaps able to make something new of me.

But I also had doubts about this search. The idea that the natural world is restorative has its roots in Romantic poetry and a good deal of recent creative non-fiction features a solitary walker in a natural landscape rediscovering or healing themselves. I began to wonder what was elided or lost when nature becomes a hospital ward for the depressed or sick at heart. Wasn't there something solipsistic in reducing place to somewhere to write the self? Must the natural world be emptied of its history and politics before it becomes home ground? This coastline was inundated with history and flooded with cultural meanings. I wanted this to become part of my story.

I'd look out from the clifftop above St Margaret's Bay, England's closest point to France, across the southern end of the Goodwin Sands to the matching cliffs of Cap Gris-Nez on the other side of the Channel. I'd remember sitting in the RHIB, water over my waist, looking avidly towards shore and the safe haven of Ramsgate harbour. I became interested in the shape of the coastline, its relation to the Sands and to the congenerous outline of France across the water, sitting there like a piece of jigsaw separated by deep time from the other piece with which it had once fitted. Nothing ever remains settled on coastlines. Around 5000 BC, Thanet, until then part of the mainland, had broken away to become an offshore island before being reabsorbed to the Kent coastline as the Wantsum Channel – separating the two – silted up in the late Middle Ages. Even today the Isle of Thanet, as it is still called, cannot be reached without crossing water. I was thinking a lot about change and reflected on how the history of human settlement along this coastline – from pre-Roman to post-Norman times – had, like the coastline itself, changed and changed again as a series of different peoples had come ashore, established themselves and intermingled with those already living there. Invasion and consolidation, erosion and renewal formed the deep history of this coast.

As seen on a map, the protruding foreland of Thanet and the inwards sweep of coastline that follows round from Ramsgate and down past Sandwich and Deal to Dover and Folkestone has the outline of a chin and neck. This geographical fantasy was suggested to me by the opening lines of an anonymous poem in Alan Sillitoe and Fay Godwin's book, *The Saxon Shore Way* (1983):

'The coast of Kent is England's chin,
Jutting out to take the punches'

Will Self has fashioned an extravagant simile to describe the Kent coast closer to London. If the British Isles is seen as a seated figure, he wrote, then the Thames is its anus, the Medway its vagina and the Hoo Peninsula 'its green and pleasant perineum'. By comparison the idea of England's chin seemed modest and plausible. Just as chins alter with age, protruding or receding, wrinkling and sagging, so too had the physiognomy of this stretch of coast. In taking its many blows, the East Kent littoral, with its White Cliffs and Dover as 'the key to England', had, through time, become figured as the bulwark of the nation, a signifier of great symbolic power. I was now living at the nation's edge, a seaboard deemed vital to national security and identity, marginal land but of defining significance.

I began to plan a book exploring these and other ideas that were forming in my mind. I'd start offshore, back on the Goodwin Sands, then come onshore and walk the coastline more systematically than I'd been doing, not to produce a point-to-point walking book, and certainly not a guidebook, but to explore what was distinctive in its history and character. I'd go underground into the caves, tunnels, mines and bunkers which the chalky terrain of this coast had enabled and where much of its history lay hidden. As I walked and began to write, Britain embarked on its fourth, then its fifth, war of the twenty-first century; the flow of refugees into Europe from countries decimated by invasion and civil war became a torrent; and the Channel-facing coast of Kent became, as it often had been in the past, a border to be defended.

Far from standing for lost causes, the east coast was becoming the defining territory of a burgeoning ethno-nationalism with hostility to incomers a focus of the many anxieties and resentments at its heart. Calais became Britain's outer border and the journey of many refugees in search of sanctuary came to a halt in 'the jungle', a term first coined by the children who experienced its squalor. Some who did manage to reach Dover were locked away in detention centres. Brexit fed the clamour for control. Border Force cutters and helicopters patrolled the Channel coast. Far-right groups rallied in Dover. A vigilante group of self-styled 'patriots' combed the shore for landings. The idea of the White Cliffs as promising refuge and hospitality, which had survived two world wars in the twentieth century, crumbled. I set out to better understand these matters by placing them in the long history of this coastline as a border defining a nation.

OFFSHORE

1

THE GOODWIN SANDS

‡

On clear days at low tide, I would stand on the seafront at Deal and gaze at the white line of waves breaking on the Goodwin Sands, blurring the long knife-edge of the horizon. In my new house I'd hung a map thickly dotted with the names, dates and types of vessel (square-rigged, lug-rigged, galleon, warship, submarine) that had foundered on the Sands. I pored over Thomas Treanor's *Heroes of the Goodwin Sands* (1892) and discovered that I'd moved into the street where this renowned chaplain to the Missions to Seaman for Deal and the Downs had lived. The house is still there and has a curious brick obelisk on its garden wall, said to be a replica of a watchtower that once stood on the roof from where Treanor could survey his Channel and North Sea parish. That's as may be, but the story added to the spell of the Sands.

Treanor described the thirty-six square miles of sand spreading along the coast from Ramsgate towards Dover that constitute the Goodwin Sands as like 'a great lobster, with its back to the east, and its claws, legs and feelers extended westwards towards Deal and the shipping in the Downs'. This 'great ship swallower', or 'widow maker' as it was also known, has ingested many centuries of shipwrecks and jetsam into its mass of deep sand sitting on chalk. In addition to a grisly toll of several thousand sailing vessels, the Sands has more recently swallowed a couple of hundred tons of iron railway sleepers, a German submarine and around sixty planes – Spitfires, Hurricanes, Messerschmitt, Dornier 17s and Junkers 88s – which came down over and around the Sands in the early years of the Second World War. Their pilots and aircrew joined the many thousands of drowned bodies already held there in a mass compressed grave.

The Goodwin Sands – six miles off the English coast – are the outer defence of mainland Kent, forming a low-lying breakwater between France and England – a rudimentary extra coastline. The relatively sheltered waters they create is the Downs, a kind of moat between the low outer wall of the Sands and the inner defence of the shoreline. In a storm, the Downs is both a haven and a danger, offering protection but also the risk of being driven onto the Sands itself or the shore, depending on the winds and tides. The Downs is where the British Navy often gathered to defend the coast from incursion or to prepare to attack the navies or coastline of other nations. It is where smugglers and those scavenging shipwrecks on the Sands confronted customs and excise men. It has been an anchorage and watering place for trading vessels waiting for favourable weather or tides to enter the Thames Estuary for home or to sail westwards down the Channel out into the Atlantic and beyond. In the mid-eighteenth century, more than four hundred merchant or warships were frequently recorded in *Lloyd's List* as anchored in the Downs which became known as 'England afloat'. Ships sailing westwards still shelter there in torrid conditions. Over many centuries the Downs has provided some of the most vivid encounters between humankind and the elements in the history of the British coastline.

The Goodwins are fundamentally unstable. Since their first scientific survey in 1840, they have been lengthening at around four hundred feet per decade. Their width is increasing too, though more slowly, and like a writhing lobster their outline is forever changing. Evidence suggests that the sand which slowly blocked the Wantsum Channel and cut Sandwich off from the sea in the late Middle Ages probably came from the Goodwins as breakaway banks migrated shorewards. As the Sands slowly extend, they could, in centuries, merge with the coast, continuing the interplay of shore, isle and sea that has characterised this stretch of the Kent coastline for millennia.

In this way the Sands slip the basic distinction between land and sea, being simultaneously of both. Such places that escape ready categorisation are often sources of fascination as well as fear. The Goodwins have been used for horseback, foot and go-kart races, treasure hunts, firework displays and a variety of sports. In Ian Fleming's *Chitty-Chitty-Bang-Bang* (1964), Caractacus Pott and his family have a picnic there. And as I well know, there is a long and dangerous tradition of playing cricket on the Sands. These are unlikely activities for an offshore setting more dangerous than anywhere else in the British Isles. Not only do the tides and winds make the Goodwins an exposed and hazardous recreation ground, but the firm sand of its dry surface goes into semi-suspension with the returning tide, becoming quicksand. The unwary or beguiled are in danger of being caught by a tide that returns at a rate of knots and drowning in the swirling currents or being sucked down into the Sands. John Conrad, as a schoolboy, was taken out to the Sands by his father, Joseph, who described the place as 'that damn piece of mother earth that had claimed so many wrecks'.

The Goodwins is also an archive of lost and buried stories. The oldest of these is to do with their legendary incarnation as the fertile isle of Lomea, said to be part of the estate of Earl Godwin. One version of this story has Lomea sinking suddenly into the sea as retribution for Godwin's crimes. Another connects its disappearance with a great storm in 1099, recorded in the Anglo-Saxon Chronicle, which resulted in the island being inundated by sand. In some versions, Godwin, returning home with his fleet, was drowned in this storm, although he is more reliably understood to have choked to death while dining with Edward the Confessor. Written accounts of this fabulous island of Lomea date from the end of the sixteenth century and were revived in the nineteenth by Charles Lyell, author of *Principles of Geology* (1830) and otherwise a great dispeller of myths. Borings conducted with the intention of erecting a lighthouse on the Sands revealed a stratum that Lyell identified as clay. He suggested that the Sands were therefore the remains of a clay island rather than the product of currents and tides. Others elaborated this to refresh the claim that the Goodwins had indeed once been Lomea, a theory that survived well into the twentieth century.

Lost islands are as old as storytelling. Nineteenth-century nautical charts and atlases contained several hundred islands that do not exist, the result of error, deception or illusion. But it is also the case that some islands, volcanic ones especially, do appear and disappear, which is one of many things that make them so imaginatively resonant. The garden island of Lomea – 'very fruitful and… [with] much pasture' – that became the hellish pit of the Goodwin Sands was a myth of paradise lost to be deployed and refashioned according to the mentality of the time. Treanor was unconvinced by the Lomea hypothesis, knowing all too well from personal experience that there were 'great natural causes in operation which account for the formation of the mighty sandbank by gradual accumulation'. He nevertheless used the Sands to elaborate his own theodicy of man's seeming insignificance and God's overarching scheme: 'The Sands reach round you for miles… In such solitudes, and with such vastness around you, of which the great lonely level stretch makes you conscious as nothing ashore can do, you realise what an atom you are in creation'.

Most stories about the Goodwin Sands involve violent storms and lawlessness. The Great Storm of 26–27 November 1703 was the worst recorded ever to hit Britain. It began with a week of gales followed by an extraordinary cyclone that burst out of the Atlantic late on the night of 26 November and moved across the country. One-fifth of the seamen of the sovereign fleet were drowned that night as were many hundreds of the crew of merchant and fishing vessels. England's cities, Bristol and London in particular, suffered terrible damage. The Eddystone Lighthouse was destroyed. Millions of trees in southern England were uprooted.

The main chronicler of these events was Daniel Defoe. *The Storm* (1704) is a founding work of reportage. Defoe advertised in several newspapers for people to send him eyewitness accounts of 'the late Dreadful Tempest' to include 'no Particulars but that they are well satisfied to be true, and to set their Names to the Observations they send'. This prospectus anticipated, by over two centuries, the Mass Observation (M-O) project initiated by Charles Madge, Tom Harrisson and Humphrey Jennings in 1937. Like M-O, Defoe sought to collect what he called 'speaking sights' to report on the condition of the nation. On an investigative trip to Kent, Defoe himself claimed to have counted seventeen thousand oaks brought down in the storm before becoming overwhelmed by the task.

The loss of warships on the Goodwin Sands that night outraged him. In his poem 'An Essay on the Late Storm' (1704), Defoe describes 'The Fatal *Goodwin*, where the Wreck of *Navies* Lyes / A thousand dying Saylors talking to the Skies', and turns on the admirals who had failed to secure the fleet in inshore harbours rather than leaving them to try and ride out the storm in the Downs. A letter from one of Defoe's respondents, J. Adams, 'from on board a ship blown out of the Downs to Norway', gave a first-hand account of the wreck of the *Sterling* (sic) *Castle*. Having been almost run down by this stricken man of war, those aboard Adams' ship watched in horror as it ran aground on the Goodwins.

It was a Sight full of terrible Particulars, to see a Ship of Eighty Guns and about Six Hundred Men in that dismal Case; she had cut away all her Masts, the Men were all in the Confusions of Death and Despair; she had neither Anchor, nor Cable, nor Boat to help her; the Sea breaking over her in a terrible Manner, that sometimes she seem'd all under Water; and they knew, as well as we that saw her, that they drove by the Tempest directly for the Goodwin, where they could expect nothing but Destruction: The Cries of the Men, and the firing of their Guns, One by One, every Half Minute for Help, terrified us in such a Manner, that I think we were half dead with the Horror of it.

The Goodwin Sands sometimes give up their secrets. In 1979 the *Stirling Castle* reappeared almost intact from a receding sandbank but was reclaimed by the Sands before it could be surveyed. It surfaced again in 1998 and stayed around longer this time, allowing for the recovery of many perfectly preserved items – a rare Dutch bronze cannon, prayer books, clay pipes, a wide-brimmed leather hat – which are now displayed in Ramsgate Maritime Museum. The embalming qualities of the Sands add to their uncanny aura and have disclosed in the case of the *Stirling Castle* a new kind of ghost ship, one that appears from the depths rather than scudding across oceans.

Defoe was venomous about the response of those on land to the plight of the many hundreds caught on the Sands on the night of the Great Storm. Rescue and salvage along the East Kent coast at this time was carried out by local craft normally engaged in fishing, smuggling and wrecking. Boats from Deal and Walmer would race each other to the Sands when ships were in trouble, cargo rather than human life their main concern. In *The Storm*, Defoe turns on the people of Deal 'for their great Barbarity in neglecting to save the Lives of abundance of poor Wretches; who having hung upon the Masts and Rigging of the Ships, or floated upon the broken Pieces of Wrecks, had gotten a Shore upon the *Goodwin Sands* when the Tide was out'. But their desperate cries for help were ignored by boats interested solely in 'Booty and… Plunder'. Deal is cursed as a pariah town to be shunned:

The Barborous Hated Name of Deal shou'd die,
Or be a Term of Infamy;
And till that's done, the Town will stand
A just Reproach to all the Land.

Deal's name was slow to recover. The town's reputation for smuggling became so bad that in 1784 the younger Pitt sent soldiers to set fire to the fishing fleet pulled up on the shore during winter. William Cobbett – like Defoe an early exponent of reportage – described Deal in *Rural Rides* (1830) as 'a most villainous place… full of filthy-looking people. Great desolation of abomination has been going on here', he wrote, inverting or misremembering the Biblical phrase. But by the time of Cobbett's visit, the town's image had begun to soften. Deal was an important naval yard during the wars with France – Nelson was based there for several years – and it figured more benignly as the setting for several nautical romances, Douglas Jerrold's famous melodrama *Black-Ey'd Susan* (1830) the best known of these.

The town's full release from Defoe's anathematisation, however, was mainly the work of its celebrated life-boatmen. During the 1860s, the recently formed Royal National Lifeboat Institution established a chain of lifeboat stations at Ramsgate, North Deal, Walmer and Kingsdown to guard the Sands and the Downs. Although competition and bad feeling between these different stations was endemic, their purpose was unequivocally to rescue rather than loot. The previously unruly relation between shore and Sands was becoming ordered and institutionalised.

One of the earliest to report and celebrate this was the prolific and ever-topical writer of stories for children, R.M. Ballantyne. *The Lifeboat: A Tale of Our Coast Heroes* (1864) is set in Deal and is a paean to 'the race of heroes by which our coasts are peopled'. The book was also a fundraiser for the Royal National Lifeboat Institution and includes statistics about the number of ships wrecked and lives lost on the Sands and advice on how to resuscitate the nearly drowned. Ballantyne had been ridiculed for including in *The Coral Island* (1858) an illustration of a shaggy coconut hanging from a palm, not realising that coconuts in their natural state have a smooth outer covering. Thereafter he often did work experience when researching his novels, spending a week on one of the floating lightships off the Goodwins when he returned to the Sands in a later novel, *The Floating Light of the Goodwin Sands* (1870).

If Ballantyne was keen to ensure his fictions were based in fact, Treanor's books about the Deal life-boatmen and the Goodwin Sands offered the real thing. Chaplain to the Missions to Seamen for twenty-six years, Treanor served merchant and foreign seamen passing through the Downs, the local coastguard stations, the nine lightships in the English Channel and the boatmen of Deal. He would hold services on board ships in the Downs embarking on long voyages and make often hazardous trips to the four Goodwin lightships and to the Galloper and Long Sand lightships anchored on the edge of the North Sea.

A large man, dressed in 'high sea boots, thick Cardigan vest and pilot jacket', his powerful voice would sound across the Downs to the accompaniment of his collapsible harmonium as he led his congregations in services on deck. A muscular Christian par excellence, Treanor embodied the qualities that he commemorated in the Deal life-boatmen, and the power of his many stories of peril and rescue on the Sands comes from having shared the experiences he celebrates.

In writing about the Sands, Ballantyne and Treanor were updating Britain's island story. For centuries the Goodwins had been a threat to the nation's seamen, to the trade upon which the nation's prosperity depended and to the naval power upon which its empire was founded. Positioned at Britain's front door, the Sands had to be tamed or at least confronted. The men and craft of the Royal National Lifeboat Institution were a modern, organised response to the threat they presented. Ballantyne and Treanor's celebration of the selfless heroism of the RNLI boatmen sees them as underwriting Britain's security and imperial reach and exemplifying the qualities which British civilisation is spreading across the world, the very antithesis of Defoe's callous looters.

I knew that I'd have to return to the Goodwin Sands. A company in Dover was running monthly trips between spring and autumn led by people who knew the Sands and timed the visits strictly in accordance with the tides. It was mid-October, the last trip of the year. The morning was dismally wet, but the winds were light. I knew that I was in safer hands this time but when I saw the boat that was to take us out, I froze. It was identical to the one in which I had been trapped, another RHIB: bright orange, thick inflatable sides, a solid bottom. Squeezing into a narrow double seat, I felt claustrophobic and shivery. There were just ten of us this morning and there was no cricket gear or TV equipment, just our backpacks, but even so, the RHIB felt dangerously overloaded.

As we moved away from our berth and towards Dover's harbour entrance, a seal popped its head up and swam beside us for a while. A good omen perhaps. Once out of the harbour we headed up the coast towards Deal, skirting the cliffs. There had been huge chalk-falls at Langdon Cliffs and St Margaret's Bay caused by the previous winter's freeze and the heavy rain that had followed in the spring. On such a dull morning the cliffs looked grubby, the vegetation on their face like stains. Where the slips had occurred, however, the sheered cliffs were a fresh dazzling white, with the appearance of a very demanding ski slope. It's falls like this which renew their whiteness.

Heading further out into the Channel, north-east towards the Goodwins, brought home how very low-lying the flat sweep of coast from Kingsdown to Thanet is. Sea and land share the same flat level. An enormous wave would carry miles inland with nothing to obstruct its course. The coastline here is a counterpart to the Sands; two inadequate bookends, two ineffectual breakwaters edging the Downs, which is a moat virtually without walls.

It had stopped raining, but the day remained grey and sullen. This time the Sands appeared first as a thin, smudgy line on the horizon. The lack of wind meant there was no splashing of waves breaking on their edge, the common signature of the Sands. We stopped just off their southern end to transfer into a small inflatable we'd been towing. Its outboard motor wouldn't start, and I began to feel queasy. Stuck again. But the sea was so calm that one of the crew managed to wade through the shallow water with a rope attached to the inflatable and pull us onto the Sands.

This time, without the distraction of a camera crew and a cricket match, I could experience just how weird it feels to come ashore in the middle of the English Channel. The tide was low and the sand under my feet was solid, hard and rutted, abrasive like roughened concrete, uncomfortable to walk on. Treanor describes it well as being 'dense' when dry but 'friable and quick' when the tide returns, and how 'if you stand still you slowly sink, feet and ankles, and gradually downwards'. As before, there were puddles, 'foxholes' as I'd learned they're called, and deeper, water-filled holes, 'swillies'. Dipping my foot into them, I had the sucking, sinking feeling I remembered from my previous trip. When the tide returns, the whole of the Goodwins becomes like this, mutating from a hard solid to something almost liquid.

The Sands were much narrower at this southern end – I was on one of the lobster's rear legs – and it was only a few hundred yards from where we landed to walk across to the opposite side facing France. Sand, water and the French shore were continuous with each other, a single surface, as if the jigsaw pieces scattered by geological time had been fitted together again. It seemed almost as if I could walk over the water to France. To the north-east, the Goodwins stretched away with no hint of the North Sea beyond like an iron-sand desert. The scene was unbounded, and I lost all sense of distance and scale.

It was dead quiet, the weather placid, the Goodwins slumbering. Treanor had described their 'deep sullen roar' but today there wasn't a murmur. Even the gulls were silent. It was hard to imagine I walked on the graves of thousands of men and ships. Treanor reports how he'd seen 'the long line of a ship's ribs swaddling down into the sands, and… the stump of the mast to which the seamen clung last year', and I'd read that even now wrecks were often visible at low tide and that a submarine occasionally re-emerged from the depths. Today nothing could have seemed less likely. Everything was calm and serene. Cosseted by a properly organised expedition, with no cricket gear to weigh me down, I wandered away from the party and lost myself in this strange setting.

Suddenly, though, I remembered the sands in Wilkie Collins' *The Moonstone* (1868), which would quiver and heave as the tide turned and an 'awful shiver' spread across their surface 'as if some spirit of terror lived and moved and shuddered in the deeps beneath'. Within an hour or so, the water would come rushing back, flooding the Goodwins from all sides and welling up from below as they disappeared beneath the surface like the ships and the bodies they have engulfed. Today, however, the timing was perfect and there was no panicked scramble for the boat when the call came to leave. Our return to Dover was as calm and smooth as the rest of the expedition. I brought away a small phial of Goodwin sand, which now sits amid several ships in bottles and beneath the map of Goodwin shipwrecks in my home on the coast.

The expressive power and resonance of the Goodwin Sands is such that they can speak to and for most periods. A plan to construct a massive lighthouse topped with a cast-iron statue of Queen Victoria surveying the Channel from a perch more than a hundred feet above the Sands was unmistakeably of its time. The Duke of Wellington, Warden of the Cinque Ports (he died in Walmer Castle), was one of its sponsors and an unsuccessful attempt to erect the monument was made in 1841. A smaller beacon was placed there instead but removed in 1850 because it had shifted 150 yards and sunk by nine feet, offering a perfect text for many Victorian sermons. During two world wars, the Goodwins were as militarised as the Kent coast they imperfectly sheltered. Mine sweepers and patrol boats protecting the Channel were based on the Sands, and torpedoed vessels were so frequently brought there for inspection and repair that the Sands became known as 'the hospital'. In the 1960s, during the years of post-war prosperity, there was a scheme to construct a deep-water port on the Sands and combine this with a two-runway airport. In 2013, a plan for a four-runway hub airport on the Goodwin Sands was one of the proposed sites considered, though wisely rejected, in the interim report on the third London airport.

The Goodwin Sands have long been a place where dreams and follies are wrecked. In 1991 the Radio Caroline vessel *Ross Revenge* ran aground there and the era of offshore pirate radio and a special moment in the history of counterculture came to an end. And several seasons after its near catastrophe on the Sands, Beltinge Cricket Club also came to an end, the older members grown too old, the younger ones departed. I have not heard of any cricket being played on the Goodwin Sands since.

ONSHORE

2

INVASION AND DEFENCE

‡

Different regions, coastlines especially, have distinctive stories they tell about themselves. For the Kent coast it has been a narrative of invasion and defence. From Roman times at least this has been a militarised seaboard, defending the shore from incursion and protecting the sea routes to the Thames Estuary. It has also been a coastline from where raids and invasions across the Channel to Europe and places beyond have been launched. The landscape itself speaks of this narrative, alternating as it does between the commanding protective heights of the White Cliffs and exposed alluvial flatland, which seems to invite incursion. Back on dry land, I set out in search of its many landmarks and ruins that speak of 'our island story' – 'This fortress built by Nature for herself'.

The twin towers of Reculver stand like an isolated chess piece in a prolonged endgame on an almost empty board. The coastline stretching south is interminably flat and straight, the low cliff on which the towers perch the only raised ground for miles. Nothing else breaks the line of the horizon. It's a mild, calm, early spring morning; the sky is milky and the towers look both evanescent and enduring, ghostly and monumental. The Romans built a fort on the site where the towers now stand, one of a chain they constructed to counter Saxon shore raids. Excavations have unearthed numerous infant skeletons, possibly ritual sacrifices, under the remains. There's an old story of a crying baby said to be often heard among the ruins, a story made more poignant by the name the towers commonly go by, the 'Two Sisters'. Ford Madox Ford, writing in 1900, recalled that until a few years previously, human bones stuck out of the hillside below the towers and that there was still a placard announcing the penalty for stealing them.

In Roman times I'd have been standing a mile inland, not on the edge of the coast. I'd have been looking across the Wantsum Channel, a wide, fast-flowing tidal river, to the Isle of Thanet a few hundred yards away on the opposite bank. In Anglo-Saxon times I'd have been standing beside a monastery, and five hundred years later a parish church. By then, the twelfth century when the towers were built, the Wantsum was silting up; Thanet was being slowly reabsorbed to the mainland; and encroachment from the sea had brought the remains of the fort, the monastery and the flourishing parish church much closer to the shoreline. Where ships had once sailed through the Wantsum Channel into the Thames Estuary, the land was closing in around them. Centuries later, the resulting marshland had been drained and enclosed as farmland. Only a vestige of the Wantsum now remains, just a trickle really, over which you step a few miles inland at the village of Sarre if walking from Thanet to Canterbury.

Meanwhile the sea had continued to nibble at the shoreline. By the beginning of the nineteenth century, the towers stood just 150 yards from the coast and the church had been abandoned. The towers were then purchased by Trinity House, the body responsible for the shipping and well-being of seafarers, as a navigational aid. Groynes were built to protect the cliff on which the towers had been left standing and the tide they faced was stemmed. Although the coastline either side of the site has continued to recede, the now safeguarded towers are forming a new headland along this coast.

Standing on the promontory beside the towers, ringed by the flat horizon, there are traces of this past everywhere. At my feet and stretching down the coast towards Minnis Bay are the groynes built by Trinity House, some constructed out of large rocks; others, much older, are wooden, rotting and covered in weed. From here they look like the remains of shipwrecks.

Directly in front of me are the hatcheries from where young oysters are transplanted to the seabed at Whitstable. A sixteenth-century writer recorded that oysters from Reculver were 'reputed as farre to passe those of Whitstaple, as Whitstaple does surmount the rest of this shyre in savorie saltnesse'. A millennium and a half earlier, the Roman poet Juvenal had noted that oysters from the 'Rutupian shore', that is the coastline around Richborough at the other end of the Wantsum, were a great delicacy. Several hundred yards beyond the oyster hatcheries, where mussel beds lie just offshore, the present-day River Wantsum cuts its narrow, diminished, meandering way inland. Once a navigable seaway, it is now the thinnest of channels, indistinguishable from the other scraggy little inland waterways that furrow ruts across this part of Thanet.

The towers themselves, lonely relics of the past, seem out of time and out of place in the world at their feet and the sea they overlook. Immediately below them are two caravan parks – 'Waterways' and 'Blue Dolphin' – deserted this morning but busy in summer. A little way inland up on the low eminence of Thanet's ridge, the vast glasshouses of Thanet Earth, the largest hydroponic low-carbon complex in Britain, are gleaming like a celestial city. Directly out to sea, the metallic turbines of the Kentish Flats Offshore Wind Farm are lined up like the monstrous Martian tripods that invade the English countryside in H.G. Wells' *The War of the Worlds* (1897). Together with its big brother, the Thanet Wind Farm, further down the coast, these wind farms have the appearance of a massed armada forming a pincer movement concentrated on the towers.

There is a curious interplay here between the crumbling fortitude and forlorn dignity of the steadfast towers, the cheerful holiday world of the caravan park and the shimmering promise of the hydroponic and wind farms. In this stand-off between a distant past and a vivid present, the towers are the most likely to survive. The wind farms are owned by the state-owned Swedish energy company, Vattenfall. Thanet Earth is underwritten by a European consortium. Both ventures depend on the backing of European finance. Most of the workforce at Thanet Earth is East European. The towers, by contrast, protected by the many tonnes of huge rocks that have been piled around the base of the natural plinth on which they stand, will continue to memorialise the rise and fall of different communities and the geographical changes the place on which they stand has undergone. They have withstood and outlasted so much and are likely to continue doing so beyond the time when the turbines of the wind farm have succumbed to the erosion of the sea or the consortia that own them have gone bust.

As I stood looking at the narrow inland course of the present-day Wantsum with these thoughts in mind, I realised I had the choice of two coastlines to walk: one present-day, the other historic. Reculver had opened up a deep topographical and historical past that pointed back down the old Wantsum Channel to Richborough, where in 43 AD the Romans had come ashore and established an imperial gateway. I decided that the present-day Kent coast could wait a bit. I would trace the contours of the old shoreline that once separated Thanet from Kent in order to explore how England's coastal border must have appeared to the Romans, Saxons and Vikings who had, in the short space of a few hundred years, fought their way ashore and transformed the cultures they found. I was also intrigued by the idea of exploring twinned coastlines, one inside the other.

The old river bed of the Wantsum Channel is threaded with narrow ditches – lodes – that drain the marshland and make it possible to farm. Lacking crossing points, and just too wide to leap across, these runnels form a latticework that make it impossible to follow a straight path. A mile or so inland, I lost the Wantsum for a while as it curved away west towards Chislet Marshes. After a tedious detour, I met it again at Sarre where it merges with the Great Stour, a rivulet swallowed by a river. I was walking through a wide polder – flat fields of reclaimed marshland – that is also a drainage basin where water from all the surrounding marshes – Chislet, Stodmarsh, Westmarsh, Sarre, Monkton and Minster – oozes into the single waterway of the River Stour at Plucks Gutter. Ten miles further on the Stour flows into the English Channel at Pegwell Bay, completing the line of the old shore.

Plucks Gutter is named after a Dutch drainage engineer, Ploeg. The Dutch were experts at draining marshland by creating ploughed ditches and the polder I'd been trudging across – the word itself is of Dutch origin – is the result of their expertise, another example of the many different cultures that have shaped this corner of England. Standing on the arched bridge over the Stour at Plucks Gutter, I tried to imagine the course and scale of the historic Wantsum Channel. Its far bank was marked by Minster, across the marshes from where I stood. The other bank lay behind me in the direction of Sturry and Canterbury. I was in the middle of the old Channel looking across terrain that, over many centuries, had slowly mutated from seaway to silted channel to salt marsh and now to arable farmland. There is no reason to think that this gradual metamorphosis is over. Centuries hence, this land will very likely have been absorbed by the sea as water levels rise and coastal erosion deepens. Boxlees Hill, Clapper Hill, and Weatherlees Hill, tiny bumps in the flat, featureless landscape I was contemplating, will become part of the seabed.

The walk had been flat and samey – miles and miles of bugger all – and I was tired. But staring across the marshes towards Minster, the land seemed to gently shake itself into life, a sensation that revived me as well. Gerard Manley Hopkins wrote in his journal for 1871 that, 'what you look hard at seems to look hard at you'. He was writing about clouds, a beautiful passage: 'The bright woolpacks that pelt before a gale... solid but not crisp, white like the white of an egg, and bloated-looking'. But here, from the bed of the old Wantsum Channel, it was history and geography that was looking hard at me. Just east of Minster is Ebbsfleet, the landing place of the Saxons in 449 and Augustine in 597. Minster is one of the oldest centres of Christianity in Britain, a rich monastic settlement and nunnery having been established there in the wake of Augustine's successful mission to convert Aethelberht and the Kingdom of Kent from paganism to Christianity. And where I am standing, here in the middle of the old Wantsum, Alfred had trapped and defeated invading Danish ships at the end of the ninth century.

In Thomas Hardy's imaginative world, the 'past-marked prospect' is invariably cheerless, reminding the gazer that unnumbered others have come this way, made the same mistakes, suffered and died. This was a setting which, for Hardy, would have suggested the bleakness of anonymity and repetition. But as the landscape quietly shrugged off its centuries, I was stirred by what it disclosed: erosion, attrition and loss, certainly, but also mutation and adaptation, recovery and renewal. From being one thing, it had become another, and another. I felt the restless cultural and geographic shape of this coastline looking back at me, speaking of change.

Richborough, at the Channel end of the Wantsum, is the reverse image of Reculver. Whereas Reculver, once a mile inland, now sits on the edge of an eroding coastline, Richborough, now more than a mile inland, was, when the Romans landed there, perched above a coastal harbour at the wide mouth of the Wantsum. On coming ashore, this huge army of forty thousand soldiers dug ditches to defend its elevated beachhead and, over the next thirty years, established a busy town and trading centre. Richborough became the imperial entrance to Roman Britain, its significance symbolised by the construction of a monumental arch, at least seventy-five feet high, which would have dominated the low-lying land and inland waterway at its foot and been visible from far out in the Channel. By the end of the third century, Saxon raids had caused Richborough to be reconfigured as a fort, one of a chain around the eastern and southern coast of Britain and twinned with Reculver in defending the entrances to the Wantsum.

Sitting on this warm spring afternoon with my back against the massive remains of the west wall of Richborough Roman Fort, I was still thinking about the mutating shape of this coastal edge, how mainland and island have separated and merged in a slow-motion pas de deux as if never quite able to commit. The interplay of past and present was even more apparent here. The wide course of the former Wantsum Channel was more obvious than at Reculver, and so too was the evidence of Roman occupation. Richborough is surrounded on three sides by the remains, eight metres high in places, of the original walls of the fort and the deep ditches that edged it. On this quiet afternoon, with the fort entirely to myself, the presence of the past became overwhelming. I could imagine myself as one of the few remaining soldiers of the last garrison just before the Romans withdrew in the early fifth century. Where there had been water was now land, but the flatness and the silence folded past and present together like a squeeze box.

Yet the view was also very different from that described by Charlotte Higgins in *Under Another Sky* (2013) just a few years earlier. When she looked out onto 'the green ocean' of the former Wantsum, 'the only vessel riding it was the giant hulk of a concrete power station' and its triple cooling towers. But very soon after, these towers, the only prominent vertical feature in the landscape between Thanet and Deal, were brought down by controlled explosion. One Sunday morning I had joined others lining a ridge between the inland villages of Woodnesborough and Ash to watch the three hundred-foot cooling towers crumble and collapse inwards. There was a festive atmosphere. Residents with a grandstand view of the event had invited friends and brought out the champagne. *This is what public hangings must have been like*, I thought. First came the sight of the collapsing towers, then the sound of the explosions rolling across the plain, next the judder of the small earthquake they caused and, finally, a dense cloud of dust. Within thirty seconds, what had been there was there no longer, like sudden death. The power of the explosion and the speed with which the towers came down silenced us all. As we dispersed, it felt like turning away from a graveside after a burial.

The power station had opened fifty years earlier, burning coal from the Kent mines. Converted to burn oil, by the end of the 1980s it was using Orimulsion, an experimental fuel described by Friends of the Earth as 'the world's dirtiest'. Acid rain from its sulphur emissions damaged crops and the paintwork of cars. It was decommissioned in 1996 after an action was successfully brought against its owners, Powergen, since when it had stood idle.

Slowly, the cooling towers came to be regarded with affection, absorbed into the local geography. With their scooped sides and spreading skirts, they had the appearance of outsized mannequin dummies. They became known as the 'Three Sisters', linking them with the 'Two Sisters' at Reculver. There were many letters of protest in the local press at their demolition. A runner described how they would guide him home. I had used them in a similar manner on long walks and cycle rides. For other letter writers they were a monument to the coal mining past of the East Kent coast, a memorial to the twelve men killed during their construction, a reminder to someone recently arrived from the north-east of the 'Angel of the North'. In the short space of half a century, they had come to articulate a complex history of social and personal meanings. And then, abruptly, they were gone.

They leave, however, a trace. A few hundred yards from where the cooling towers stood is a solitary wind turbine, long since decommissioned, a foundling left by the 'Three Sisters'. So flat is the terrain that this too is visible from miles around, looming suddenly where one is surprised to find it, a waymark allowing the walker or cyclist to orient themselves, a scout sent by the many hundreds of its kind massed out in the Channel. For me this lone turbine has become a cross or cairn marking the life and death of the 'Three Sisters'.

Towers in our time express global corporate power, those other 'Twin Towers' for example, or the thrusting 'iconic' buildings of the City of London. The monumental arch built at Richborough, probably the largest of its kind in the Roman Empire, was an analogous expression of imperial power. It was a symbolic and ceremonial entrance to the colony of Britannia, purposely aligned with the Roman road, Watling Street, which led directly to Canterbury and London. Large amounts of marble from the imperial quarries at Carrara went into its making. It was designed to awe and astonish from both land and sea. Shops and roads were built around its base as the town became a major port and trading centre.

In times of conflict, by one means or another, towers and monumental arches come down. The 'falling towers' of T.S. Eliot's 'The Waste Land' spell the end of a succession of civilisations: 'Jerusalem Athens Alexandria' and so on. When Richborough became a fort in response to Saxon raiders, the monumental arch was broken up and used in the construction of the massive walls against which I leant my back. Some of the Carrara marble was burnt to provide lime for the concrete, and other chunks of it are still visible in parts of the outer wall. The passive grandeur of the arch had failed to keep 'the hooded hordes' at bay and so the Romans prepared for siege, the beginning of the end of their presence on this coast.

A millennium and a half later, Richborough was again on the front line, not this time as the site of the biggest-ever invasion of Britain but of the biggest-ever outwards movement of munitions. During the First World War, an extensive port was built at Richborough on the River Stour's winding route from Sandwich to Pegwell Bay, from where munitions and other supplies were produced and shipped to the Western Front and barges were constructed and ferried across the Channel to be used on the canals of northern France. This was the Royal Engineers' secret harbour, although the scale of the project meant that it can't have been a secret for very long. It had a permanent complement of twenty thousand men. Hundreds of women were brought in each day from surrounding towns and villages to sort the shell casings ('empties') that had come back from the battlefields of France and Belgium. A chain of camps – Stonar, Haig, Kitchener and Cowan – spreading down the coast towards Sandwich and inland towards the remains of the old fort was constructed to house this vast army, comparable in size to the former Roman army of occupation. Among these buildings was a detention block. Graffiti from its walls is displayed in Ramsgate Maritime Museum, including a sharp little poem, 'World and Earth', that I'd like to see included in future anthologies of First World War poetry:

> *In the eye of the public there is no war*
> *But, in the eye of the soldier there is no earth.*
>
> Sapper A.S. Grant, RE

Access to Richborough Port during the First World War was closed to anyone who didn't work there, and it's still a difficult place to explore. The easiest way is by boat from Sandwich, winding down the Stour past Stonar Lake and the former Pfizer pharmaceutical site, on past Bloody Point and Stonar Cut to Shell Ness and out into Pegwell Bay. River trips to see the seals – money back if none sighted – are run by Captain Colin, Sandwich Harbour Master, an old river dog with skin like weathered tarpaulin, whose eloquent commentary turns apparently featureless marshland into a place of vivid historical and natural life. He tells us that the Stour is now half the width it was a hundred years ago when the Royal Engineers had finished their diggings, a result of the same process of silting that centuries earlier had turned Sandwich from a thriving coastal port to a backwater. He points out pyramid-shaped anti-tank stoneworks sunk in the riverbank, the deserted wharves and the remains of the old train ferry berth on the mud flats at the mouth of the Stour. It was from here in the closing months of the war that hundreds of tanks, ambulance trains, vehicles and mounted guns were shipped on ro-ro train ferries to Calais and Dunkirk. It was also from here that hay to feed the 800,000 horses in the front line was sent. Purpose-built electric hay presses built at the port made this possible.

The captain patrols the river as if it is his own. A solitary figure on a decaying jetty – a riverine hermit in a shaky smallholding of rotting wood – is an intruder, a trespasser on the banks of the captain's watery estate. He is happier with the natural life, pointing out oystercatchers, sandpipers, egrets, cormorants and the handsomely coloured shelduck with its chestnut, black and white plumage, dark green head and neck, and red feet and bill, which he particularly admires. There is a pod of common seals, twenty or so, at the estuarine end of the river so there will be no refunds today. Grey seals, common on the Goodwin Sands, are less frequent here – porpoises, now very rare.

But the river cruise kept me at a distance from the history I was exploring. I wanted to get closer to what's left of Richborough Port. So, I walked in from the north, through Pegwell Bay nature reserve, where a board lists the recent sightings of birds: whimbrel, knot, bar-tailed godwit, sanderling, turnstone, brent goose, Sandwich tern, avocet, fulmer cuckoo, lesser whitethroat, turtle dove, grasshopper warbler, wheatear, greenshank, Cetti's warbler, bullfinch, peregrine, marsh harrier, water rail, hobby, Arctic tern, common sandpiper. An extraordinary variety to be found in such a small habitat within earshot of the busy road between Ramsgate and Sandwich. It's a fresh, clear May morning. Looking back across Sandwich Bay to the village of Pegwell sitting in a niche in the cliffs, I can make out the Belle Vue Tavern where the young Queen Victoria used to enjoy shrimp teas made from those small crustaceans caught in the bay and turned into paste at the local Banger's Factory.

From the path I can see lines of truncated concrete pillars sticking up like amputated thumbs, designed to protect the coast from incursion, and beyond them the remains of the train ferry berth. I leave the path and head towards the river. The tall rushes bend forwards under my weight and provide a springy footing over the boggy ground. It's like walking on a trampoline. The steel base of the ferry berth is in good condition and its wooden framework, though slowly rotting, is still massive: huge timbers, the sawn-off end of one showing its arrested growth rings, embedded with thick, rusted bolts. During the First World War, this wharf would have stood out over the water but now, because of silting, it's at least fifty yards inland. As soon as the Royal Engineers departed at the end of the war, the Stour would have begun to close up.

I set out upriver, walking once again on what had been river bed, pausing to examine a torpedo casing lying like a discarded cigar and, further on, a wide, shallow-bottomed boat silted into its mooring. Squeezing round a rusting spiked barrier, I clambered onto the long, straight line of the New Wharf, built during three hectic months in 1916. Engineering workshops manufacturing cranes, high-speed lighters and seaplanes had spread along this wharf during the First World War, and during the Second, the Mulberry portable harbours used for rapid offloading during the D-Day invasion of Normandy.

The tide had turned and was racing inland towards Sandwich, reclaiming much of the bed over which I'd walked. I scrambled up a bank and cut back into the nature reserve. It was somewhere here that all those local women had sifted empties, but I could find no trace of the old salvage depot.

I wanted to get to the other side of the Stour, to see the workings of the port from its opposite bank. The present-day coastline of Thanet would still have to wait. I walked in from Sandwich Bay Estate several miles further down the coast. The estate is an arid collection of early twentieth-century mansions where the Conservative politician Jonathan Aitken had his constituency home, the 'White House', at the time of his conviction and imprisonment for perjury in 1999. The original name of Aitken's former mansion, 'Bleak House', better describes the house and its desolate setting. This stretch of coast is so flat that whenever you look up from the stones as you walk, the view is always the same and its sparse features never seem to get any closer. Another phrase of Hardy's came to mind – 'oppressive horizontality'. Trudging on, I recalled the Damascene experience described by Aitken as he walked this same shore: 'Suddenly, yet quietly, I became aware of someone else's presence on the beach. For a moment I thought I heard the crunch of footsteps on the shingle behind me, but when I turned round, no one was there... But someone was – I sensed them strongly at first, then overwhelmingly... "Slow down," said a gentle voice somewhere inside my head'. Wittingly or not, Aitken is mawkishly echoing a line of 'The Waste Land': 'Who is that third who walks always beside you?'.

At the mouth of the Stour there was plastic everywhere – barrels, bottles, boxes, bags, traffic cones – great mounds of the stuff, the onshore equivalent of the Great Pacific Garbage Patch. Even the rotting remains of dead fish looked like plastic. Mixed in with it all were some old clothes and a couple of suitcases. I looked around half-expecting to see a body. Amid this wasteland of rubbish, the old ro-ro ferry berth just across the river seemed lost and insignificant, overwhelmed by plastic. I was comforted, though, to see the lone wind turbine poking up above the treeline of the bank opposite. As usual, it wasn't where I'd have expected it to be.

Plodding back over the stones the way I had come, I passed a man and woman fishing with the arid plain behind them. They looked up at me, surprised. "Where have you come from?" they asked.

"New Zealand," I replied. It seemed the only answer. On past Prince's links and Royal St George's Golf Club, at the edge of the deathly Sandwich Bay Estate, a Royal Navy bomb disposal van drew up alongside a waiting coastal rescue van, and together they disappeared down the road towards the

Stour Estuary. Another unexploded shell or mine must have been washed ashore.

South to Deal, the coast path fringes the grassed-over dunes of the Royal Cinque Ports Golf Club dotted with pill boxes in which golfers now shelter when squalls hit. Then, suddenly, I'm among Tudor defences: three fortresses – the Castles of the Downs – built by Henry VIII in the late 1530s to meet the threat of French invasion. The first of these, Sandown Castle, has all but gone, its remains now part of the sea wall. Deal and Walmer castles, however, have been carefully preserved. Together, these comrades-in-arms formed a line of low-lying fortresses built along a three-mile stretch to provide an enfilade of canon fire to defend the long, flat exposed shore, more neck than chin, either side of Deal. Water and land here form a single level; the sea is always brimming at your feet. If an army could have walked on water, it would have been able to cross the Channel and step ashore as if it were traversing a plain. In such a setting, castles had to hug the ground.

Deal carries scars and memories of recent conflict too. Dogfights and bombing raids over the town were frequent in the early years of the Second World War. One Saturday morning in 1941, German planes bombed Deal's high street, killing fifteen people and injuring many others. Gordon Blain's greengrocers, Percy Comfort's ironmonger's shop, and St George's Hall were among the buildings destroyed or badly damaged. The body of Raymond Files, fourteen-year-old son of a local coal miner, was never recovered. His father led a group of fellow miners who dug down into the basement of Gordon Blain's where Raymond was believed to have been when the bombs hit. Their search was made more horrifying because bits of meat from the butcher's shop next door were mixed in with the debris. And in 1989 the Provisional IRA set off a time bomb at the Royal Marines School of Music, killing eleven young bandsmen and injuring over twenty more. There's a memorial bandstand to the dead on Walmer Green divided into eleven sections, one for each of those killed, their names on plaques surrounding the base.

Several miles south of Deal, at Kingsdown, chalk uplands replace the shingle bank that has shaped the coast since Richborough, and the signs and traces of war intensify. Below me as I climb the steps from the pebble beach to the clifftop path is a prohibited area marked 'Ministry of Defence Property'. This was a Royal Marines rifle range during the Second World War and today men with metal detectors are foraging for ordnance. Up past yet another golf course, the fourth since Richborough, my mobile pings with a message welcoming me to France (at the end of Deal pier, I've been welcomed to Belgium).

It's a clear afternoon; the Channel has its familiar creamy skirting where the chalk has turned the water cloudy; the sea is calm and the cliffs commanding. Cap Gris-Nez stands out so clearly across the water I can make out the striations on its white cliffs. The prospect along the cliff edge focuses on a soaring needle-like column rising from its highest point. This four-sided tapering obelisk is a memorial to the Dover Patrol 1914–1919 whose naval vessels and fishing trawlers patrolled Dover Strait and accompanied troops across the Channel. The memorial is grey, grave and immense, a monolith looking across the Channel to its sister, the French Dover Patrol Monument at Cap Gris-Nez. Now, in 2020, there's a new Dover Patrol clearing the Channel of refugees attempting hazardous journeys in small boats in their bid for sanctuary.

There are reminders of other wars. At the entrance to The Pines Garden at St Margaret's Bay, where soldiers were housed and trained during the Napoleonic period, a proudly defiant sign announces that since the invasions of Julius Caesar and William of Normandy, 'the waters of the English Channel have not been crossed by an invader'. As if to underline this boast, a large bronze sculpture of Winston Churchill by Oscar Nemon stands in the gardens. St Margaret's Bay is Britain's closest point to Europe, and Nemon's sculpture – heavily clad, torso leaning forward, chin out, bulky sloping shoulders, embedded up to its shins in the base of the sculpture as if it has sprung naturally from the chalk and soil of the cliffs – holds the valley that runs up to South Foreland Lighthouse. Churchill was Nemon's great subject, the most famous of his many sculptures being the one in the Members' Lobby of the House of Commons – an unlikely career swerve for someone whose early work included a sculpture of Freud's dog, Topsy.

During the Second World War, the coastline from St Margaret's Bay to Dover became known as 'Hellfire Corner', and there's a permanent 'Hellfire Corner' exhibition of photos, maps and memorabilia in The Pines Garden Museum. In its midst, sitting behind a desk, cigar in hand, is a life-size dummy of Churchill. Standing alongside it as I examined a map of the Second World War batteries around St Margaret's Bay, I was startled by the wretched thing bursting into speech – "We shall fight on the beaches…" it growled – sending my blood pressure into orbit.

In a memorandum to the Chiefs of Staff Committee in August 1940,

Churchill had emphasised the need to maintain 'superior artillery positions on the Dover promontory… to fight for command of the Straits by artillery, to destroy the enemy batteries, and fortify our own'. He instigated and supervised the installation of a pair of fourteen-inch guns – 'Winnie' and 'Pooh' – above St Margaret's Bay, capable of hitting German artillery dug in across the Channel. He also insisted that both guns had wooden dummies placed near them as decoys. There was a story during the war that a German aircraft had dropped a wooden bomb on one of them. These big guns, and another pair established at nearby Wanstone Farm, just inland from South Foreland Lighthouse, were the showpiece of Britain's coastal artillery, but their painfully slow rate of fire severely limited their effectiveness. In Dover, it was thought their main effect was to draw even more fire on the shell-flattened town.

The last and most useless of the Kentish Heavies, 'Bruce', was an experimental hyper-velocity gun with a sixty-foot barrel designed to fire 256lb shells over seventy miles. Only it didn't. During testing – its shells were fired north towards the Essex coast – the pressure of firing caused many of the shells to explode in mid-flight and the barrel was found to have a life of only twenty-eight rounds. It was never used operationally. I came across the remains of 'Bruce' – a rubble of broken slabs and bricks strewn over a moss-covered concrete emplacement at the side of an overgrown path – where it had sat out the last two years of the war looking at the tip of the Dover Patrol Monument peeping above the treeline.

The smaller artillery of the South Foreland Battery was far more effective in preventing German shipping passage through the Strait of Dover. Four medium-range guns were installed at the top of the valley behind South Foreland Lighthouse, out of sight of the coast but assisted by the new technology of radar. These worked in tandem with the Fan Bay Battery, just south towards Dover, which operated as a flash battery to turn enemy ships into the path of the larger ones. The valley is littered with remains of the South Foreland Battery: a humped, earth-covered magazine, the concrete base of several gun emplacements, the old power house and the site of the Fortress Plotting Room which, because chalk shines so brightly from the air, was painted in creosote to camouflage it from enemy planes. Even now you can smell the creosote when it rains.

I have little interest in military history and I'm all but a pacifist, by which I mean that I can't entirely rule out a situation in which I might possibly feel compelled to fight. At my age the question has become theoretical, but in the late 1960s, at a time when New Zealand had troops fighting in Vietnam and I was called up for military training, I went before a conscientious objection tribunal. Rather to my surprise, the tribunal accepted my case, which was a more limited one than it would be now. So why was I hunting out gun emplacements, old magazines and other remains of war?

The history of the coastline I was exploring made this unavoidable, of course, but there were other influences in play. The commemorations of the First World War, which were becoming insistent as I began my walking, were complicating my quest to establish home ground. Memory is always full of forgetting and I was vexed by the highly selective narratives that were being offered. There was little acknowledgement, for example, of the hundreds of non-British pilots – Polish, Czech, Irish and Palestinian, as well as those from Commonwealth countries – who'd flown in the Battle of Britain, some of them no doubt in engagements above where I walked. The BBC's headline four-parter *The World at War*, written and presented by Jeremy Paxman, had, in defiance of its title, represented the First World War as almost exclusively a conflict between Britain and Germany. The Eastern Front was ignored, as was the colonial contribution. Even David Olusoga's *The World's War: Forgotten Soldiers of Empire*, a welcome corrective to Paxman's blimpish narrative, had taken no account of the contribution of Britain's settler colonies.

New Zealand, whose forces included Maori and Pacific Island soldiers, had suffered more losses proportionate to population than any other allied country. For me this was not a matter of pride. New Zealand had a long tradition of militarism stretching back to before the First World War. It was a nation partly founded on military victory over Maori, which had offered troops to rescue General Gordon from Khartoum and contributed enthusiastically to the Anglo-Boer War. New Zealand's twentieth-century history had been shaped by its participation in two world wars, its national identity often said to have been formed at Gallipoli and the Somme. These wars had also helped shape the history of my own family. My grandfather had served in France and Germany in the First World War, my father in the Middle East and Italy in the Second. My grandmother's brother, Alf Trevarthen, was killed at Armentières near the French-Belgian border early in 1917. There's nothing remarkable about any of this, except that these men, like all those other New Zealanders who fought in two world wars, had been sent twelve thousand miles across the world to fight – a uniquely long distance between the front line and the home front. There was no home leave for my father or grandfather. And Alf Trevarthen had never returned.

My parents became engaged on the eve of my father's departure for war a mere three weeks after they had met. They didn't see each other again for four years and they married within a month of my father's return at the end of the war, pretty much strangers to each other. My mother, a nurse, had worked in male orthopaedic wards during the Second World War, looking after severely injured soldiers who had been shipped home. Born in 1946, I grew up hearing my father's war stories, dressing up in bits and pieces of uniform he'd brought home, undergoing compulsory military training at my local state high school and marching with other school cadets to the town cenotaph on Anzac Day.

Hunting out the leavings of war along the Kent coast made all this vivid again. The gun emplacements, diggings and other deposits of armed conflict I'd found were full of stories, not just of this coastline or of Paxman's First World War but of my background and upbringing, of where I'd come from as well as where I'd come to. It brought me up against much that I disliked, a century of war whose shadow I had grown up in, whose legacies I had resisted but which were common to where I was raised and where I now lived. This wasn't quite the home ground I'd been looking for, but it was becoming part of my story.

Fan Bay, between South Foreland and Dover, couldn't be called anything else; it names itself. It spreads out from a narrow base near sea level, widening as it extends upwards and inwards to form a kind of vast amphitheatre, as if the cliffs were a massive block of ice cream that had been scooped out with an enormous ladle. From the bottom, the handle of the fan as it were, a thin line of steps has been cut into the face of the slope, dangling above me like a rope ladder. I started counting the steps as I climbed but somewhere around the ninetieth, arithmetic became impossible. The footholds were stable enough, but the gradient triggered a vertiginous fear of toppling over backwards and I went up the last bit on all fours, like a fly on a wall.

Halfway up, and just off the line of the steps, were signs of recent diggings. These I discovered were where the Fan Bay sound mirrors were buried. Sound mirrors were an early warning device installed around the south coast during and after the First World War to detect the approach of enemy aircraft, a kind of aural radar. Known as 'whisper dishes' or 'listening ears', they were concave concrete constructions often with a microphone at the front to amplify the sound caught by the dish. Quaint as they now seem, and by the 1930s the increasing speed of aircraft made them of limited use, the system of linked stations for plotting aircraft movements they formed was adopted by the radar system that replaced it. At the top of the bay, just above the path skirting the curving edge of the fan, I came across the recently excavated and heavily bolted entrance to the deep shelter where the men of the Fan Bay Battery had lived. The National Trust was excavating the site to uncover the mirrors and open the deep shelter to the public. Volunteers were needed and I made a mental note to find out more.

South Foreland Lighthouse, like many of the vertical structures around this coastline, has a twin, North Foreland Lighthouse which sits on a headland at the eastern extremity of Thanet. South Foreland is an early Victorian lighthouse. It was the first ever to use an electric light, was where Marconi received the first ship-to-shore message and became an early radar station during the Second World War. It dominates its setting like a small castle, looking out boldly across the Channel, down on Dover harbour to the south and lining up with the Dover Patrol Monument to the north. Onshore lighthouses patrol the threshold between land and sea from a secure roost. South Foreland, painted a dazzling white, seems like an outcrop, an upthrust from the White Cliffs on which it sits. It is short and spunky, doughty and confident, contrasting in colour and scale with the mass of sombre dark stone that is Dover's Norman castle.

Dover Castle and the town itself express perfectly the narrative of invasion and defence that defines this coastline, epitomising some supposed essence of national character and occupying a special place in the national imagination. Operation Dynamo – the Dunkirk evacuation – was planned in the underground network below Dover Castle. Camera crews stood on the cliffs above Dover in 1940 to film dogfights between British and German aircraft over Dover Strait, just as the inhabitants of the town had gathered in the same place to watch English sloops and frigates repel the Spanish Armada. The Romans are said to have been beaten off from landing at Dover and forced to come ashore further up the coast at Ebbsfleet. The site of Dover Castle was a fortification as far back as the Iron Age, a clifftop telling a narrative of defence against incursion that pushes back to before the beginning of recorded history.

Dover's Western Heights, across the valley from the castle, is part of the same story. Looking up at the Heights from the town, the view is simply of a grassy, partly wooded hillside with a low skirting of houses, but they secrete an extraordinary network of fortifications begun in the late eighteenth century to counter the threat of invasion by the French. William Cobbett, visiting in 1823 during the course of his rural rides, described miles of hillside 'hollowed like a honeycomb'. Deriding the wasted expenditure of what he saw – 'what reason had you to suppose that the *French would ever come to this hill* to attack it' – he was struck by the novel appearance of these fortifications:

> This is, perhaps, the only set of fortification in the world ever framed for mere *hiding*. There is no appearance of any intention to annoy an enemy. It is a parcel of holes made in a hill, to hide Englishmen from Frenchmen.

In fact, Cobbett was noting a decisive change in the method of defending the town and the coastline. The development of large cannon able to batter castle walls from a distance meant that the idea of a castle with high walls, towers, ramparts and keeps dominating a coastline, which Dover Castle so splendidly represents, was now being replaced by systems out of sight of an invading force.

Cobbett would have been even more contemptuous if he had lived to see the defensive diggings prompted by the fear of Napoleon III in mid-century: an additional four miles of moats up to sixty feet deep, a maze of tunnels, drawbridges, stairs, wells and firing positions. Citadel Battery, looking out across the Channel over Shakespeare Cliff, dates from this time and was used again in both world wars. Its guns were removed in the 1950s, but the battery is otherwise intact. The Citadel itself, a grim moated fort at the top of the Western Heights, became the Dover Immigration Removal Centre in 2003.

I wish I liked Dover more. It's been dumped on often enough and hardly needs me to do it again. Thousands of buildings in the town were destroyed or badly damaged during the Second World War. More than two hundred civilians were killed and hundreds seriously injured. Whereas London was defended by air patrols over the Weald of Kent, East Kent was too close to the Channel for early warning and at the start of the war was defended only by out-of-date anti-aircraft guns. The big guns that replaced them drew more fire on the town. Many of the inhabitants spent the war living in caves in the cliffs and in tunnels beneath the town.

Dropping down into Dover from the White Cliffs in 1951, Alan Sillitoe wrote:

> *One would not think that the war had ended six years ago. These ruins look so established… Nearly all the hotels along the harbour front are empty and roofless, and behind their facades the extensive area of desolation is like the Cocteau Hades in the film* Orphee.
>
> *Back in empty streets, doors have been ripped from shops and houses… The framework of windows has fallen out, strips of wood in shreds. Wet wood hangs through a ragged hole that opens into the basement. Blue paper faded and rotten clings to walls streaked with damp.*
>
> *Where there is no house at all, rectangular holes are like open tombs. Bushes with ten years' growth sprout their branches up basement walls and make tracks across the pavement.*

Even now, Dover struggles to repair the damage. I'd set out walking from South Foreland Lighthouse to Dover. Coming down from the cliffs along a wide, grassy slope where a railway line constructed for the building of Dover harbour once ran, a kestrel windhovers above me, almost within arm's reach, poised and taut like a tightrope walker securing their balance. This moment of mastery passes quickly as I drop down onto Marine Parade, following in Sillitoe's footsteps past the scrubby edge-land between the road and the cliffs where a derelict hoarding offers 'Roadside Development Opportunity'. A dead seagull lies on the footpath beside a bus stop. Two letters have dropped off the sign announcing, 'Dover Leisure Ce t e'. A thin, pale, pimply teenager is on his mobile and as I go past, I hear him whining, "Piss off, Mum."

I'm embarrassed by these vignettes and by the voyeuristic melancholy they provoke. But there's no escaping it. Dover is a run-down place, the deprivation of those who live here, and those who are detained here, obscured by its historical characterisation as 'the key to England' and by its White Cliffs as the symbolic expression of proud island nationhood.

Above all, Dover is a place of transit, the busiest passenger port in the world. Looking down from the cliffs above the Eastern Docks, I had watched hundreds of container lorries gathered to board the next ferry, edging forward one by one towards the stern doors in perfect straight lines, as if a child at play was patiently moving them across a carpet. My path led down to a short dead-end road, Athol Terrace, beside the roundabout where the traffic turns into the docks under a large sign, 'Departures'. A narrow terrace house overlooking the sign is called 'Matthew Arnold House' and has a yellowing copy of the first stanza of 'Dover Beach' in a small glass case by its front door. The only 'melancholy, long, withdrawing roar' today is that of container trucks in low gear making their steep ascent into the ferries. The poignancy of the public display of this poem is heightened by the modest private nature of its presentation, hand-typed lines on fading paper in the kind of cabinet that might be used for a parish notice. The plangent cadences of Arnold's poem suit the mood of the place.

The ceaseless comings and goings of the ferry port capture and fix the essence of the stretch of coastline around which I've been walking. Dover is not the end point of my journey – I want to push on to Folkestone and I've yet to walk around the Thanet coast – but it epitomises so much of what I've come to think of as England's chin. Arrival and departure, the constant mixing of peoples, 'an atlas of tongues' as Auden puts it in his poem 'Dover', a militarised coast from at least as far back as the Iron Age, a shoreline to defend, a border to keep strangers out. Invasion and defence, however, is not the whole story of this coastline. Invasion had often led to settlement, to human and cultural exchange. Not all landings have been hostile, nor have all those who came ashore been invaders. Some have sought refuge and hospitality and not infrequently this had been offered.

At Reculver, at the start of my walk, I could hear only the gulls and my tinnitus. It was a place of quiet, of settled history. Richborough had felt similar, the present receding into the past. Dover, by contrast, is a place of noise, of the here and now: the grating rumble of the traffic, voices from the docks floating up to the Heights, the streets full of those departing and those arriving or seeking entry, 'an atlas of tongues' indeed as I walk by. The present mood, however, is to keep others out. The hesitant optimism of Daljit Nagra's poem 'Look We Have Coming to Dover!', in which the displaced 'hutched in a Bedford van' cross 'the vast crumble of scummed cliffs' to a freedom of sorts, 'babbling our lingoes, flecked by the chalk of Britannia', has also crumbled. Hundreds of the unwelcome have been detained on top of the White Cliffs in the Citadel, a fortification built against the threat of invasion. Dover's chalk cliffs have become a barrier rather than a sign of welcome.

3

PAINTING AND WRITING THE SANDS

‡

I was back at Minnis Bay to resume my postponed walk around the present-day Thanet coastline, thoughts of invasion and defence still uppermost. But the broad sandy bay with its rows of colourful beach huts, blue with yellow doors, took my thoughts elsewhere. Thanet had once been famous for its sands, a shoreline of pleasure and amusement, London's Riviera, and on this bright, warm early summer morning, I could see why. At West Bay, the beach huts had been thrown open and were being cleaned out and spruced up in readiness for the coming season. Unfolded deck chairs were freshening in the sun. Here the huts were all painted cream with doors of alternating blue and yellow and numbered. A terrace of them ran from 1–41, like a little Lego street. At St Mildred's Bay, dog walkers were sitting chatting at tables outside Millie's Beach Bar. St Mildred's had been a Royal Navy seaplane base during the First World War, but today this barely interested me. I'd come to the seaside.

The pleasure of the seaside was a Victorian discovery, and I was approaching its *locus classicus*, the celebrated sands of Margate, Broadstairs and Ramsgate. At first the trippers and holidaymakers had come from London to Thanet by steamer, a scene memorably described by Dickens in one of his *Sketches by Boz* (1836), 'The Tuggses at Ramsgate':

> *The sun was shining brightly; the sea, dancing to its own music, rolled merrily in; crowds of people promenaded to and fro; young ladies tittered; old ladies talked; nurse maids displayed their charms to the greatest possible advantage; and their little charges ran up and down, and to and fro, and in and out, under the feet and between the legs of the assembled concourse...*

There were old gentlemen, trying to make out objects through long telescopes; and young ones making objects of themselves in open shirt-collars; ladies, carrying about portable chairs, and portable chairs carrying about invalids; parties waiting on the pier for parties who had come by the steam-boat; and nothing was to be heard but talking, laughing, welcoming and merriment.

The Tuggses, a family of grocers, have come into money and decided that the first step to gentility must be to get out of London. Gravesend is 'low', Margate is 'worse… nobody there but tradespeople', but Ramsgate is 'just the place' for those seeking social advancement.

Within a decade, however, the railway had come to Ramsgate, the journey from London Bridge taking just three hours, not much longer than weekend rail travel from London to Thanet can sometimes take now. Ramsgate, according to the *Illustrated Times*, had retained its upwardly mobile reputation:

Margate is vulgar… it does not wear gloves, never dresses before dinner… It is Ramsgate smoking a clay pipe, with its coat and boots off… They consume a vast quantity of ardent spirits at Margate, but at Ramsgate bottled beer is in fashion, and at Broadstairs a bottle of sherry will last for three dinners… Margate is a 'jolly place'… Ramsgate is a 'genteel town', and Broadstairs a 'dull and grand watering place'.

Victoria often stayed at Ramsgate, first as Princess, then as Queen, and many luminaries of the period did the same: Coleridge, Darwin, Wilkie Collins, among others. Some returned year after year; Coleridge, for example, an enthusiastic swimmer who loved cold water and took full advantage of the bathing machines that facilitated the growing enthusiasm for sea bathing.

From the mid-1850s, Karl Marx and his family often joined the throng. During the heat and drought of 1857, when the Thames was mostly sewage and the stench unbearable, Marx packed his family off to Ramsgate while he worked feverishly to complete *A Contribution to the Critique of Political Economy* (1859). In the early 1860s, his three daughters would spend July enjoying a fashionable life on the Sands that was outside their social world and beyond their means in London. In 1871, after an intense five days at the first International Congress after the Paris Commune, Marx and Engels took off for Ramsgate. Joining their families, they walked the cliffs, bathed, watched fire-eaters and Punch and Judy shows on the beach, and drank heavily. The last family visit to Ramsgate was in the summer of 1880. Marx and his wife, Jenny, took a cottage where they were joined by their daughters and the ever-supportive Engels. A New York journalist came to Ramsgate to interview Marx and described meeting his family on Ramsgate Sands: 'It was a delightful party – about ten in all – the father of the two young wives, who were happy with their children, and the grandmother of the children, rich in the joysomeness and serenity of her wifely nature'. In fact, Jenny Marx was already ill with the cancer from which she died the following year.

Between 1830 and 1880, Ramsgate was transformed, as Rosemary Hill puts it, from an 'elegant Georgian resort' to a 'brisk Victorian town', and became a vivid example of the social mixing that seaside holidaying produced. The Sands nibbled away at social hierarchy and physical inhibition. Jane Carlyle described as 'truly disgusting' the 'cries of prawn shrimps and lollipops', the sight of men 'in their shirt sleeves… at open windows eating shrimps', and the din of brass bands, 'Ethiopian bands', bagpipes and French horn players. *The Observer* was disturbed by the sight of women on the beach staring at naked male bathers. Spencer Thomson's *Health Resorts of Britain* (1860) spoke for those who were repelled by the seaside scrum:

> *How is it, that amid the well-bred visitors of Ramsgate… both modesty and manners seem to be left at their lodgings, so that bathers on the one hand, and the lines of lookers-on on the other, some with opera-glasses or telescopes, seem to have no more sense of decency than so many South-Sea Islanders. Ramsgate, it is true, has not Margate's extensive sands, but, surely, it is not obliged to huddle its bathers together; and loungers are, surely, not compelled by the paucity of the walks, to select the very immediate vicinity of the machines for their walk at high noon.*

Others, though, were fascinated by the scene, none more so than William Powell Frith, whose painting *Ramsgate Sands* (also known as *Life at the Seaside*, the title inscribed on its back) was the first of several large canvases of the Victorian crowd for which he is celebrated. These are now regarded as utterly conventional, forgetting that it was Frith himself who established the genre. In *Ramsgate Sands* he transferred domestic painting from the home to the beach and expanded its social range to include, in his words, 'all sorts and conditions of men', as well as of women and children. An early response to this painting dismissing it as 'a piece of vulgar Cockney business unworthy of

being represented even in an illustrated paper' – the critical equivalent of Jane Carlyle's disgust – makes clearer its innovatory character.

Frith worked at his panorama for several years, finishing it just in time to be exhibited at the Royal Academy in 1854. Its fame was secured when it was seen and bought by Queen Victoria and voted the Royal Academy exhibition's picture of the year. Victoria knew Ramsgate well and had taken her first-ever dip in the sea there just a few years earlier, as she described in her journal:

> *Drove down to the beach with my maid and went into the bathing machines, where I undressed and bathed in the sea (for the first time in my life), a very nice bathing woman attended me. I thought it delightful till I put my head under water, when I thought I should be stifled.*

The mixing of 'all sorts and conditions' on the Sands could hardly be better illustrated.

The seaside is often said to have replaced the inland spa as the Victorian health resort of choice – Ramsgate became known as 'the lung of London' – but its keynote was pleasure. Frith's painting catches this. Children are paddling, floating boats, digging in the sand; adults recline in the sun, reading, knitting, observing the scene. In the middle distance there is a group of blacked-up musicians (Jane Carlyle's 'Ethiopian band'), gipsies selling shell-work, a man with a hare, a Punch and Judy theatre, donkeys, bathing machines. The background is filled with the buildings and monuments of the town: the harbour clock tower, the Obelisk, the castellated Pier Castle, Albion House where Princess Victoria convalesced in 1835 while recovering from serious illness, the elegant terraces of Wellington Crescent and a massive slab of white cliff.

The Sands are packed with holidaymakers. As one commentator remarked, Frith, who had twelve children by his wife and seven by his mistress, seems to have liked crowds. The painting has perhaps a hundred figures. It is full of visual clues and nudges more readily understood in 1854 than today. A reviewer identified the group of father, mother and three daughters at the centre of the painting as attempting to recreate the high wall of their garden in Peckham here on the Sands, where 'they turn their backs upon everybody, living as it were within a ring fence'. The family behind them is 'not so well fed'. Another reviewer elaborated the different stories implied by each group: the affectionate old couple sharing an umbrella whereas all the other umbrellas are held singly; the ladies with their 'uglies' (a fashionable novelty attached to the brim of the bonnet to shade the face from the glare) casting sly glances on all that is passing around them; the widow still in mourning clothes being courted even though the funeral baked meats have scarcely cooled. One could imagine the Marx family among the crowd.

Bathing machines with their modesty hoods (an overhang to allow women to enter the water unseen) are lined up to the right of the canvas but we see no one in the water. In fact, the painting has almost no water, just a token line of inert froth (frith-froth) skirting its front. Ruskin's assertion that it was impossible to paint the sea has rarely been better illustrated. The point of view is from the water looking onto the beach, as if Frith is paddling with his back to the sea. The behaviour of the crowd on the Sands is its focus.

The steep uplift of the cliff and buildings at its back block out most of the sky and give the painting its concentrated and packed feel. Only telescopes point out of the picture. On the far left, a man peers intently across the sweep of the scene, his wife looking up at him from her newspaper as if to ask what he thinks he's about. At the water's edge, a little girl has her telescope pointed in the same direction. On the far right, hard up against the edge of the painting, a man (said to be a portrait of Frith himself) has some kind of glass to his eye and is turned in the same direction. The implication is they are all looking at bathers coming down the steps of the machines into the water. The man on the left might well be behaving like a South-Sea islander; the little girl's gaze is presumably more innocent; and Frith's own curiosity is curious. Here, as elsewhere, the painting is simultaneously cautious and incautious, intriguing and titillating us while ensuring the disorder we glimpse is contained. Bathing is respectable, even the Queen has swum here, and modesty is protected by hoods, yet it attracts an undesirable male gaze which is corrected though not cancelled by other more innocent or artistic perspectives. We see both modesty hoods and telescopic peeping.

There is a further aspect to the painting: national health and vigour, an implied patriotism. Here, in Ramsgate, we are on the edge of what Gladstone was to call 'the silver streak': 'Happy England!... that this wise dispensation of Providence has cut her off, by that streak of silver sea... from the dangers... which attend upon the local neighbourhood of the continental nations'. This echoes John of Gaunt's 'scept'red isle' speech in *Richard II*: 'This precious stone set in the silver sea / Which serves it in the office of a wall'. Wellington Crescent, elegant and secure above the Sands in Frith's painting, fronted the parade ground where the Duke of Wellington's troops had trained to fight Napoleon.

The street that cuts into the crescent is still called Plains of Waterloo. Albion House, where Victoria had convalesced, is the apex of the painting. Those on the Sands below, no matter what their class, are healthy and happy. The Kent coast, crucial to the protection of the nation, is at ease; its people fit, relaxed and at one with each other – the microcosm of a harmonious and confident nation.

I make the twenty-minute walk from Ramsgate harbour to Pegwell Bay, up the steep slope of the Royal Parade and the Paragon where Darwin stayed in 1850 and which Van Gogh sketched while living in Ramsgate in the mid-1870s; then along West Cliff Promenade, past Pugin's Victorian Gothic house, The Grange, with a panoramic view of the great bite of Sandwich Bay down the coast to Kingsdown where the White Cliffs resume their protection of the shore. The Tuggses make this journey, somewhat uncomfortably, by donkey. One donkey heads directly for a public house, another for a hedge, and the third grinds the leg of its rider against a brick wall. But the donkeys are eventually brought under control and the family settles down at Pegwell Bay to a plate of large shrimps, crusty bread and bottled ale. The tavern is unnamed but very likely the Belle Vue where Queen Victoria so enjoyed her shrimp teas.

Not only were Pegwell Bay shrimps a delicacy enjoyed by both the young Victoria and the Tuggses, but they also played a part in the origin of species debates of the early nineteenth century. William Paley's *Natural Theology* (1802), which argued that God's benevolent design was immanent in nature, had used the example of shrimps to demonstrate that organic life was able to take pleasure in its own existence:

> *Walking by the sea side, in a calm evening, upon a sandy shore, and with an ebbing tide, I have frequently remarked the appearance of a dark cloud, or, rather, very thick mist, hanging over the edge of the water… When this cloud came to be examined, it proved to be… filled with young shrimps, in the act of bounding into the air from the shallow margin of the water, or from the wet sand. If any motion of a mute animal could express delight, it was this… Suppose, then, what I have no doubt of, each individual of this number to be in a state of positive enjoyment, what a sum, collectively, of gratification and pleasure have we here before our view?*

This exuberant and fantastical Lewis Carroll world of leaping shrimps was linked to Pegwell Bay in *A Guide to the Coast of Kent*, published in 1859, the same year as Darwin's *The Origin of Species*.

The pleasures of Pegwell Bay, shared equally by the Tuggses at noon and shrimps in the evening, are entirely missing from William Dyce's great painting, *Pegwell Bay, Kent – a Recollection of October 5th 1858*, exhibited at the Royal Academy in 1860. Another beach scene, it could hardly be more different from Frith's. *Ramsgate Sands* is set in midsummer; the beach is bathed in sun and the tide is cheerfully up. In *Pegwell Bay* it is autumn, early evening; the sun is sinking, and the tide is mournfully out. Like Frith, Dyce uses members of his family as models for some of the figures in his canvas, including a self-portrait, but the effect is entirely different. Instead of the packed, tight-knit and interacting family groupings of *Ramsgate Sands*, Dyce's painting is sparsely populated. Its figures searching the shoreline for shells and fossils or fishing for shrimps shrink to insignificance as the perspective lengthens and deepens, leaving them isolated in the brooding elegiac scene of rock pools left by the ebbing tide, the pebbled beach, the crumbling, time-worn cliff and the dizzying presence of Donati's Comet in the sky.

Donati's Comet had appeared in the sky over England at 6pm on 5 October 1858, which explains the precision of Dyce's title. The first comet ever to be photographed, it had last been seen several centuries before Christ and is next due to return in 3811. I'm not sure of the date. Dyce has placed it above Pegwell Bay close to where Augustine came ashore in 597, bringing Christianity to England. The compression of precise time with the seeming infinite in Dyce's painting is vertiginous. On the one hand, it has the accuracy of a railway timetable, on the other, vistas of time beyond human grasp. By combining the astronomical (the comet) with the geological (the cliffs, the rocks and the shore), the human figures are dwarfed and diminished, anomic and lost. They look superimposed, like pasted-on cut-outs that might easily be unpeeled or come unstuck from the canvas, leaving the comet and the shoreline to itself.

But the everyday life of the painting is not necessarily extinguished by the immensities of time and space that surround it. Its conspicuous irony, that no one in the painting apart from the figure of the artist (Dyce himself) notices the comet in the sky, is not, I think, at the expense of the shell gatherers and shrimp fishers on the shore. Their failure to notice the comet is not because they're unobservant or self-deceiving. They're immersed in the natural world at their feet and there's nothing in the painting to suggest this is obtuse or consolatory.

For Dyce, and for others at mid-century, the seashore was a place of enquiry. This was nicely expressed by J.G. Francis in his *Beach-rambles in search of sea-side pebbles and crystals* (1859):

The point where sea and land meet is the critical point for all observers of Nature. Here the disciple of geology should serve his apprenticeship… those who desire to note epochs in the flight of Time, and to set up way-marks in the Earth's chronology, must study the line of the sea-coast, the ancient and the modern, for here, if anywhere, the dial-plate is uncovered, and the shadow of the gnomon may be traced through some seconds of the enormous day which has witnessed the existence of the heavens and the earth.

Dyce's figures are finding the infinite at their feet, where the shells, rocks and organic life of the seashore have become the mid-century equivalent of Blake's grain of sand.

As Thomas Hardy so dramatically expressed, the vastly lengthened and expanded perspectives of time and space opened up by Victorian science could seem to make all human endeavour petty and futile, but in the present moment nothing was more important than the individual life. In his novel *A Pair of Blue Eyes* (1873), Henry Knight, hanging by his fingertips from the Cliff Without a Name, comes eyeball to eyeball with a Trilobite, a strange meeting that reduces him to a being of infinitesimal significance. But for Elfride, standing on top of the cliff, nothing could be more important than saving the life of the man she loves. She removes her many undergarments and twists them like the strands of a cord to form a linen rope long enough to rescue Knight from this literal cliffhanger, putting her own life in peril as she pulls him to the top of the cliff.

Dyce's painting contains both Hardy's points of view – the universal and the particular – holding different possibilities together within its frame. Perhaps, as it has been argued, it shows the sun setting on humankind. Comets, from Aristotle and Ptolemy to the group-suicide of the Heaven's Gate sect, have commonly been seen as portents of disaster. But in Giotto's fresco *Adoration of the Magi*, for example, the comet represents the Star of Bethlehem. In striking contrast to *Ramsgate Sands*, *Pegwell Bay* is a work of multiple perspectives, a kind of Victorian Cubism with different and competing stories to tell.

These two paintings, therefore, seem like chalk and cheese. Frith celebrates the pleasures of the seaside, the modern world of the Sands. Dyce unsettles these benign amusements, replacing them with the forebodings suggested by the presence of the pre-human past. Frith fills his canvas with the enjoyment of the present; Dyce shrinks the present by placing it in the context of deep time and troubling futurity. Two beaches along this strip of coast, no more than half a mile apart, allow, even encourage, such apparently contrasting points of view.

But, in fact, both these scenes are equally modern, building sandcastles and staring at bathers through telescopes being no more of their time than searching for shells and fossils and gazing at comets. The study of cliffs, rocks, fossils, shells, jumping shrimps and indeed comets was also one of the pleasures of the age. We know that the Darwin and Dyce families spent time exploring Pegwell Bay while holidaying in Ramsgate and I would be surprised if the Friths had not done likewise. Even the foolish Tuggses, having finished their shrimps, 'went down the steep wooden steps… which led to the bottom of the cliff; and looked at the crabs, and the seaweed, and the eels'.

In July 2020, Comet NEOWISE appeared in the skies over the Kent coast. I watched for it from the seafront at Deal on three successive evenings but the weather each night was cloudy; a large, bleary moon hazed my view; and the night lights of Dover and the Thanet towns were polluting the skies. The country was in lockdown. The first wave of deaths from Covid-19 had subsided, but a further wave seemed inevitable, and the health of the country was in the hands of a shambles of a leader. Each day was the same as the last; my sense of chronology had dissolved; and space had shrunk to the confines of my house, garden and the seafront. The comet, by contrast, was a time-lord unbounded by space. Frustrated at having missed seeing NEOWISE, and thinking of Dyce's painting, I was drawn back to Pegwell Bay.

I parked at Queen Victoria's favourite tavern, walked along a path towards Ebbsfleet and found some steps that took me down to the shore. The weedy chalk-strewn shoreline was deserted except for a man in a bright red jumper digging a row of deep holes at the tide's edge. I turned and looked up at the cliff-face. The physical scene of Dyce's painting was unmistakeable. The cliffs have eroded a bit more to form new caves and arches, and fresh chunks of fallen chalk, worn smooth and rounded by the tide to look rather like ostrich eggs, lie at their base. But the overall shape and texture of the cliffs is unchanged. One of the features of the painting is its careful delineation of the layers of different geological periods that have gone into their making: the Cretaceous period when the chalk was formed; the Paleogene when sand and mud were deposited on the chalk; and the Ice Age when woolly mammoths and rhinos roamed across Kent. I could still see these layers distinctly marked in the cliff-face.

The man in the red jumper made me think of the vivid red shawl worn by one of the women in the foreground of Dyce's painting, the only splash of colour in the otherwise subdued tones of his canvas. Here on the same littoral where the Saxons had come ashore, where Augustine had landed, and in the lee of chalk cliffs revealing a remote geological past, he could have been any man at any time searching for shellfish. When Henry Knight is confronted by the Trilobite, Hardy writes: 'Time closed up like a fan before him'. For me, immersed in the setting of Dyce's painting, time opened out like a fan. I thought again of NEOWISE, formed around the time of the birth of our solar system 4.6 billion years ago, next due to return in 8786 and, because of light pollution, likely to be the last comet ever visible to the naked eye.

Suddenly, I heard the sound of a helicopter, and there it was, coming up the coastline from Dover, hovering over the precise spot where Dyce had positioned Donati's Comet. I blinked. The superimposition of the helicopter on the scene Dyce had so scrupulously rendered seemed miraculous. But the spell was quickly broken. This had become, after all, a familiar sight as Border Force scoured the coast searching for little boats bringing refugees across the Channel. The helicopter circled above Ramsgate and then moved out of the frame, turning back down the coast towards Dover as it continued its hunt.

In late October 1921, T.S. Eliot's wife, Vivien, wrote to a friend: 'Margate is rather queer, and we don't dislike it'. Margate's reputation had changed little since the Tuggses had turned up their noses at it. A.G. Bradley's chapter on Margate in *England's Outpost, the Country of the Kentish Cinque Ports*, also published in 1921, begins: 'What is there to be said about Margate, save that to the uttermost bounds of the kingdom, and beyond them, it stands for everything which makes holidays for the Cockney proletariat?'. Vivien's litotic appreciation was about the most one could expect.

The Eliots had come to Margate a couple of weeks earlier. Eliot's health had collapsed and he'd been diagnosed as suffering from a nervous complaint. His employer, Lloyds Bank, gave him three months leave on full pay, and Eliot departed London for the Kent coast and a period of solitary rest with instructions not to exercise his mind. Vivien settled him into The Albemarle Hotel in Cliftonville and stayed with him until the end of October before returning to London, leaving him a mandolin to relax with. Cliftonville, as Bradley acknowledges, was different from 'the frankly popular atmosphere of the old town', a suburb of 'fine hotels... smart terraces... well laid-out gardens, tennis courts, and bowling greens... the resort of the prosperous and wealthy'. Back in London, Vivien wrote of her husband to Bertrand Russell: 'He is at present at Margate, of all cheerful spots! But he seems to like it!', which indeed he did. Eliot's letters from Margate are contented, and when he left the town on 12 November, he told Richard Aldington he was 'very sorry' to be departing. Ever since his childhood holidays near the fishing village of Gloucester on the New England seaboard, he had enjoyed being at the coast.

Eliot's idea of relaxing at Margate was to resume writing 'The Waste Land'. The first two parts – 'The Burial of the Dead' and 'A Game of Chess' – had been completed in the early months of 1921, but then the combined effects of his job at the bank, his effort in setting up *The Criterion* and Vivien's frequent ill-health meant that his work had stalled. The arrival of his family in London in June, which involved Eliot and Vivien handing over their flat and moving to cramped quarters, triggered his collapse.

But in Margate, he started to write again. At the end of his stay, he wrote to his friend, Sydney Schiff:

> *I have done a rough draft of part of part III, but do not know whether it will do, and must wait for Vivien's opinion as to whether it is printable. I have done this while sitting in a shelter on the front – as I am out all day except when taking rest. But I have written only some 50 lines, and have read nothing, literally – I sketch the people, after a fashion, and practise scales on the mandolin.*

These 'fifty lines' became lines 259–311 of the published poem, from 'O City, city, I can sometimes hear' to the final single-word line of part III, 'burning'. As Lawrence Rainey, the Sherlock of 'Waste Land' studies, established, they were written during the fortnight after Vivien had left and Eliot was on his own. Each day he would take the tram down the hill from Cliftonville and along the seafront to the Nayland Rock shelter, near Margate's renowned Royal Sea Bathing Hospital, where he sat and worked on his poem. Before this, under Vivien's watchful eye, he had only been allowed to strum his mandolin.

Part III, 'The Fire Sermon', moves from 'the brown fog' of London, Eliot's 'Unreal City' with its shabby and unlovely inhabitants, down the Thames to Margate Sands where the third Thames daughter sings of how she can connect nothing with nothing. The dominant mood of ennui or *aboulie*, the word Eliot used to describe his own condition at the time, has momentarily lightened just before the poem departs London for Margate, heralded by Ferdinand's line from *The Tempest*: 'This music crept by me upon the waters', and those immediately following, 'Beside a public bar in Lower Thames Street / The pleasant whining of a mandolin / And a clatter and a chatter from within / Where fishmen lounge at noon'. The promise of this music takes us out onto the river with its red sails and the beating oars of Elizabeth and Leicester's gilded barge. The wind freshens, just as it has in the opening section, 'The Burial of the Dead', when the music of Wagner first breathes life into the poem.

But this mood is negated by the songs of the three Thames daughters telling of seduction, abandonment and loss, first on the river at Richmond, then at Moorgate, finally at Margate.

"On Margate Sands.
I can connect
Nothing with nothing.
The broken fingernails of dirty hands.
My people humble people who expect
Nothing."
La la

Just as in *The Tempest*, Ferdinand is misled by Ariel's beguiling song of invitation to 'come unto these yellow sands', Margate is a sandy shore and Thanet a silted isle without magic. The voices of these lost women fade as those of St Augustine and the Buddha's Fire Sermon conclude the section.

What would Eliot have seen as he sat upon the shore? Looking over his right shoulder, he'd have noticed the new signs announcing the renaming of Margate's famed seafront amusement park as 'Dreamland'. By 1921, after the war years, Margate was re-establishing itself as a seaside resort and holidaymakers were beginning to return. Jeremy Millar has remarked that the line 'On Margate Sands.', with its full stop, is like the inscription on a postcard sent home from a seaside holiday; a plain legend beneath a photograph. Eliot was, after all, on a holiday of sorts. I'm intrigued by the thought, suggested to me by the poet Simon Smith, of Eliot looking across those yellow sands and drafting these sections of 'The Waste Land' with Dreamland at his back. The names are a dead rhyme, a dull echo of each other. Could the title of Eliot's poem have been prompted by a funfair?

Less speculatively, Eliot was in the midst of a recent war zone. Margate had been frequently bombed from the air during the First World War, by Zeppelin and aeroplane. In Margate Museum, I found a wall map of where the bombs had hit that shows a concentration of attacks around Cliftonville, where Eliot was staying, and the Royal Sea Bathing Hospital near the shelter where Eliot sat and wrote. Photos from the time show a fortified seafront and clifftop with barbed wire entanglements and other defences. There would have been bomb-damaged buildings everywhere in 1921, and local memory of the many deaths from the war and the Spanish Flu that followed must still have been fresh.

David Seabrook, musing on Eliot in Margate, described 'The Waste Land' as 'a poem which appeared to have been blown to bits and put together again at high speed' but not 'quite right'. Eliot had been deeply affected by the war. As Wyndham Lewis wrote:

> *He was an American who was in flight from the same thing that kept Pound over here, and with what had he been delected (sic), as soon as he had firmly settled himself upon this side of the water? The spectacle of Europe committing suicide – just that.*

In a letter to his cousin, Eleanor Hinkley, in 1917, Eliot described 'life over here' as like living in a Dostoevsky rather than a Jane Austen novel. London in those years would have been strangely without young men except for the unfit, the wounded and those in uniform on home leave. The ghostly final section of 'The Burial of the Dead', with its shallow-buried corpse and zombie-like crowd flowing over London Bridge, bears the impress of this time, prompting the line, 'I had not thought death had undone so many'. In the following section, 'A Game of Chess', the 'demobbed' scene is explicitly about the war, and the lines 'I think we are in rats' alley / Where the dead men lost their bones' refers to a trench in the Somme. David Seabrook also notes that National Poppy Day in memory of the war dead was inaugurated on 11 November 1921, the day before Eliot left Margate for London en route to Lausanne where he was to undergo treatment.

Looking out to sea often reminds me that I come from somewhere else. On the seafront at Margate, Eliot might also have been reminded of his position as an outsider. In his early years in Britain, he sometimes signed himself 'Metoikos', a metic, a resident alien. Although Eliot had done his best to ape the English, sometimes prompting Bloomsbury mirth as he did so, United States neutrality for much of the war left him in uncertain territory. His several unsuccessful attempts during 1918 to enlist with the United States forces added to this feeling of marginality. Although disinclined to return home, Eliot was not yet at home where he lived. This classic condition of the migrant, caught between two worlds, correlated with other more psychic divisions in Eliot at this time, most obviously that between the barren and inauthentic material world into which the figures in 'The Waste Land' are locked and an unrealisable world of the spirit or the imagination that might offer transfiguration or transcendence.

Sitting each day in his shelter looking out across the Sands and the tidal bathing pool (like the shelter, still there), over the water towards France and Belgium, from where the sounds of fighting had been audible in Thanet several years earlier, Margate became a correlative of the fragmented state of Eliot's mind and of the world – a blighted place in which voices sing of disconnection and loss. Like the other broken worlds of the poem, Margate is desolate and deathly, the Sands a desert not a beach, Thanet as *thanatos*.

At the beginning of the Second World War, John Betjeman enlisted Thanet into a construction of nationhood that looked back to Frith rather than Eliot. His poem 'Margate 1940', first published in *The Listener*, is focused on Cliftonville. The setting is the Queen's Highcliffe, a fashionable hotel overlooking the Queen's Promenade along which its habitués and those of the nearby Grand would take the air. It is early evening, 'As soft over Cliftonville languished the light / Down Harold Road, Norfolk Road, into the night'. Tables are being laid for dinner. The guests are in their rooms washing 'the sand from their legs' and preparing for 'an evening of dancing and cards'. The sound of the sea drowns the roar of the trams. All this for six stanzas, and then the last:

> *Beside the Queen's Highcliffe now rank grows the vetch,*
> *Now dark is the terrace, a storm-battered stretch;*
> *And I think, as the fairy-lit sights I recall,*
> *It is those we are fighting for, foremost of all.*

It is war again; the lights have gone out. Margate embodies a way of life that has become dimmed and threatened and must be saved.

In 1943, Betjeman gave a radio talk in which he listed the things that for him England stood for: 'the Church of England, eccentric incumbents, oil-lit churches, Women's Institutes, modest village inns... the noise of mowing machines ['Margate' opens with lawns being cut]... local newspapers, local auctions... local concerts, a visit to the cinema, branch line trains, light railways, leaning on gates and looking across fields'.

In *Notes Towards the Definition of Culture* (1948), Eliot made his own list of what the national culture might include: 'Derby Day, Henley Regatta, Cowes, the twelfth of August, a cup final, the dog races, the pin table, the dart board, Wensleydale cheese, boiled cabbage cut into sections, beetroot in vinegar, nineteenth-century Gothic churches and the music of Elgar'. I'm struck by how much it has in common with Betjeman's list, both of them nostalgic and rather complacent even then. An alternative list at the time could have included trade unions, left book clubs, surrealist exhibitions, censorship, the means test and so on.

Both lists are much narrower than they seem, and neither shows any recognition that culture is defined by difference and dissent as well as by what is held in common. Otherwise it begins to echo the ugly nationalism of the Citizen in James Joyce's *Ulysses* (1920): 'we want no more strangers in our house' – a remark directed at the Jewish Leopold Bloom. On the day I was writing this, an opinion poll had found that one-third of the United Kingdom population saw Islam as a threat to the 'British way of life'. Properly understood, any definition of the 'British way of life' today must now include Islam. I think of the narrowness of Eliot and Betjeman's list as expressing, among other things, the unease of the outsider and a consequent desire to be assimilated.

Neither man had a straightforward relation to English culture. Each in his own way felt himself a stranger, with nationality and war an issue for both men. Betjeman's family name had been spelt with a double 'n', suggesting German origins, and at Highgate Junior School he was taunted: 'Betjeman's a German spy – / Shoot him down and let him die'. His parents quarrelled over the spelling of the family name, his mother dropping the second 'n', his father retaining it.

Eliot and Betjeman might seem an unlikely pairing, and in terms of poetics, they certainly are. But back in 1916, Betjeman, aged ten, had thrust a sheaf of his poems onto 'the American Master', hoping to impress T.S. Eliot, who was for a few months teaching at Betjeman's school. It's curious to think of 'the American Master' and the boy with a German name, each a kind of alien, in the same prep school in 1916. Their acquaintance was renewed in the 1930s when Eliot, now at Faber, wanted to republish Betjeman's first collection, *Mount Zion* (1932). Friendship followed, with regular exchanges of letters, shared outings and Betjeman writing for Eliot's *The Criterion*. When Eliot turned sixty, Betjeman wrote that his soul 'travels in the same carriage as mine, the dear old rumbling Church of England which is high, low and broad at once', but again less inclusive than this makes it sound.

As late as 1976, Betjeman wrote: 'I have a terrible guilt about not having any right to be in this country', which partly explains the fogeyish persona he cultivated and his lifelong investment in a certain kind of Victorian Englishness. And in time, Eliot became so much the Englishman that his American birth and upbringing are sometimes overlooked.

By the second half of the twentieth century, Eliot's picture of Margate had prevailed. Dreams had soured, as shown in Lindsay Anderson's 1953 documentary *O Dreamland.* This twelve-minute black-and-white film, without commentary but with a soundtrack of noise and music, follows bank holiday day trippers at Dreamland as they move around the sideshows and other amusements. It's a tawdry and joyless world with none of the fun of the fair. The crowd looks on impassively at a recreation of the execution of the convicted Russian spy Julius Rosenberg who, together with his wife, Ethel, had gone to the electric chair just a few months earlier, during that 'queer sultry summer' as Sylvia Plath describes it in the opening sentence of *The Bell Jar* (1963). A jukebox repeats the song 'I Believe' – a sardonic backing – while a puppet clown laughs maniacally at the absurdity of the crowd and the spectacle it has come to view. The final shot is of the lights of Dreamland seen from above as Anderson's camera pulls away – in relief, one feels. Only after dark, with the reality of the scene lost in the glow of its famous illuminations, Betjeman's 'fairy lights', can anything be salvaged from the grotesquery of this crowd-puller.

It was also in the 1950s that the Queen's Highcliffe hotel, together with the Grand, was bought by Billy Butlin and later demolished to make way for flats. So too was Eliot's Albemarle. Soon, holidaymakers no longer came to Margate or anywhere else in Thanet, and the hotels and guest houses that survived became cheap bedsits for local homeless people, for those that boroughs elsewhere couldn't or wouldn't house, and for poor migrants seeking refuge and more tolerable lives.

Today the relation of Cliftonville to the old town has been reversed. A.G.

Bradley's 'resort of the prosperous and wealthy' has crumbled and decayed to become a scrap land of run-down housing inhabited by the displaced, many of them Roma from the Czech Republic and Slovakia, declining numbers of long-term local residents who are being squeezed out and a small number of much better-off incomers from London who have paid less than the price of a garage in the capital for their houses. In Harold Road and Norfolk Road, it is now the inhabitants who are languishing rather than the mellow light of Betjeman's poem. The term 'multiple occupancy' hardly describes the crowded, dilapidated Edwardian terraces. The fronts of these houses are thick with Thanet District Council wheelie bins, each indicating a separate tenancy. I counted more than twenty such bins outside a double-fronted house, larger than the rest, as I walked up Harold Road. Any remaining space out front was filled with ruined mattresses, a jettisoned fridge, a wrecked Dyson hoover – the debris of poverty and transience. On another house, someone had chalked: 'Please Dont I Love You!'.

There's a reminder of Thanet's Victorian past at the head of Harold and Norfolk Roads, a pair of modern gated blocks of flats named Dickens Court and Darwin Court. Dickens Court is where Eliot's Albemarle stood. Harold and Norfolk Roads run parallel to each other away from the seafront and into Northdown Road, a most curious shopping centre. I was struck, at first, by the number of animal rescue centres: 'Cats in Crisis', 'Pet Rescue and Home Centre', 'Pet Clinic – the Frank and Ethel Fright Centre', whose syntax I paused to unscramble. Having just walked up Norfolk Road, I was thinking it was the inhabitants who needed rescuing. But 'Reptile Advisory Centre' left me staring. A run of second-hand furniture shops was more predictable. Alongside one of them was a vaping centre – 'Love Life Love Your Lungs' – a far cry from Thanet as the lung of London.

The reversed fortunes of the two parts of Margate pivot on the Turner Contemporary gallery which opened in 2011. Turner Contemporary sits at the beginning of Margate's stone pier below Fort Hill (site of an Iron Age settlement) which once separated the smart set of Cliftonville from the working class of the old town but is now the border between a regenerating town and a decayed suburb. Consecutive exhibitions in 2012 elaborated, so it seemed to me, on Margate's cultural myths. Hamish Fulton's exhibition 'Walk' featured a short film documenting a crowd of 198 people walking slowly and silently around the Margate tidal bathing pool. I bought a limited edition print of one of its frames because the long line of evenly spaced walkers on the seawards wall of the pool brought to mind Madame Sosostris's lines from the opening section of 'The Waste Land', 'The Burial of the Dead': 'Fear death by water. / I see crowds of people, walking round in a ring'. This is the pool in front of the shelter where Eliot drafted 'The Fire Sermon'. Zombie-like, Fulton's walkers and Eliot's seemed to mimic each other.

The following exhibition at Turner Contemporary was Tracey Emin's 'She Lay Down Deep Beneath the Sea'. Emin is Margate's most famous daughter. The title work of the exhibition is of a reclining female figure imagined, as Emin herself describes it, 'lying under a great weight, represented by the sea and symbolising change, time, and loss'. I am now so immersed in the shape and contours of the East Kent coastline that I see it everywhere. Fancifully no doubt, the spread legs of this figure seem to trace its outline from Thanet around into Sandwich Bay and beyond to Deal. When Emin likens her work in this exhibition to cave drawings, I think of the chalk caves of this coastline with their drawings and graffiti which she grew up around. The dominant mood of the exhibition was separation and loss, though not without tenderness. Emin has always been open about the personal basis of her work, and her well-known remark about Margate as a place where 'you don't lose your virginity... you have it broken into' echoes the plaint of the Thames daughters. Born in London and moving to Margate as a child, Emin made the same journey from the unreal city to the Thanet coast.

Turner Contemporary has redefined Margate. Many of its early visitors had never been to a gallery before and its weekend crowds still look rather different from those at Tate Modern. A small village of independent shops and cafes has grown up around the Old Town Hall in Market Place. The gallery, designed by David Chipperfield, is across the road from Mrs Booth's lodging house where Turner frequently stayed in the 1830s and '40s, and the view from inside the gallery is the one that he would have seen. Turner's Margate and Emin's have been very neatly brought together, connecting then and now. And Emin, single-handedly as it were, has recently spent several million pounds in buying an old Edwardian bathhouse, mortuary and children's nursery to establish an arts school and artistic centre, the TKE Studios (her initials). Fittingly, this latter-day Thames daughter has been made a freewoman of the town.

From outside, the Turner Contemporary gallery is stylishly modest, unlike those iconic buildings described by the architect writer Rowan Moore as 'whooshy', which towns and cities seeking to rebrand themselves feel they cannot be without. In shape and in colour, it is like a stylised extension of the chalk cliffs that drop down from Fort Hill to Margate Sands. The roof tilts upwards away from the land towards the sea, creating a cliff-edge effect and maximising the offshore view from inside the gallery. The building is blue-white-milky; from Eliot's shelter at the other end of the Sands, it looks a bit like an iceberg with a few windows. Whatever the vantage point – approaching Margate down Canterbury Road from the west, from the seafront, the end of the pier looking back to the Sands, or from Fort Hill looking down – the gallery is conspicuous without dominating. It has become Margate's landmark without having to soar hundreds of feet above the town. It fits its setting without boasting, content to leave the brutalist concrete tower block of Arlington House, a protected 1950's building, to hog the skyline. From inside looking out, it is as if you're at sea, as if the walls of the building have been cut away. The galleries are spacious and acoustically subtle so that the sound of conversations drifts upwards, leaving a murmur like the receding tide in its wake.

In mid-March 1866, Karl Marx departed London for Margate, plagued with carbuncles, tired after a combative gathering of the International Workingmen's Association and under pressure to complete the long-overdue first volume of *Capital* (1867). He took lodgings in the centre of Margate overlooking the Sands and began a rigorous cure of intensive walking and sea bathing. In a letter, he described himself as a 'walking stick, running up and down the whole day, and keeping my mind in that state of nothingness which Buddhism considers the climax of human bliss'.

Marx and Eliot both went to Margate for their health and 'freedom of mind' while wrestling with the works they are most famous for. Marx was resting (if a seventeen-mile walk to Canterbury during this time can be so described) when he should have been writing; Eliot writing when he should have been resting. Both were preoccupied with nothingness, Marx more vigorously and sceptically than Eliot who turned away from Margate Sands to the Buddha's Fire Sermon at the end of the third part of 'The Waste Land'.

The symmetries are seductive. Both men were migrants, cash-strapped, with stressful domestic lives, and each enjoyed the support of a patron and friend – Engels in Marx's case, Ezra Pound in Eliot's. Engels was the more secure benefactor but both he and Pound facilitated, in *Capital* and 'The Waste Land', two of the most profound and influential visions of modernity, as too in its way did Margate. The differences between the two men are obvious.

Marx looked to the emancipation of humankind whereas Eliot, at the time of 'The Waste Land', could see only defeat. But the vision of the contemporary world in each man's writing is deeply pessimistic. Alienation – of people from the world, their labour, their bodies, their souls – is the keynote. Marx saw capitalism as sucking the life out of labouring bodies, reducing workers to automatons, mere 'hands' as Dickens described them in *Hard Times*. Eliot's poem is peopled by zombies, the living dead and the undead.

For Marx, there was a way out of the prison house of industrial capitalism. When the New York journalist who described the Marx family enjoying the pleasures of the Sands asked Marx what he thought was 'the final law of being', Marx looked at the choppy sea and the crowded beach and replied, "Struggle!" 'The Waste Land', on the other hand, is a world drained of struggle, its voices passive and defeated. There is no dialectic at work in Eliot's vision, no agent of historical change to transform a material world that repelled both men. That said, Marx's view of the actual working class could sound a bit like Eliot's in the pub scene at the end of the second part of 'The Waste Land', where the controlling voice sardonically intones 'Good night, ladies… good night, sweet ladies' as time is called, echoing Ophelia (and anticipating Lou Reed). The weather in Margate when Marx stayed there that March was terrible: 'As if it had been made especially to order for the COCKNEYS who have invaded this place for the Easter holiday', he wrote.

I walk to Marx's lodgings in Margate – formerly 5 Lansell's Place, now 10 Albert Terrace – an end-of-terrace house adjacent to the clock tower built to commemorate Victoria's Golden Jubilee in 1887. The house looks directly across the Sands to Eliot's shelter and the tidal bathing pool, from where the view cuts back across the Sands to Turner Contemporary. These three points form a large triangle with Marx's lodgings at the inland apex, structuring and arranging how I've come to see Margate's seafront. How would these two migrants, Marx and Eliot, view the town today? The straitened condition of its displaced and dispossessed would be grist to the mill of any updated Marxist analysis of global capitalism and its depredations. And in Cliftonville, if not from his shelter, Eliot would find material enough to confirm almost a century later the vision of 'The Waste Land'.

Yet, as elsewhere around this coastline, change is constant; loss is often mitigated by recovery; and, as I was soon to discover, even waste can be put to good use. The most recent stage in the rejuvenescence of Margate's old town has been the reopening of a carefully restored Dreamland, including its famous wooden roller coaster. Marx, who enjoyed Punch and Judy shows on the Sands, and Eliot, who loved music hall and admired Groucho Marx, might at least have been cheered by the return of what was originally a Victorian amusement park to share the seafront with a gallery inspired by Turner, whose genius responded best to sea and coast and who painted so intensively around the Thanet shore. The world of Frith's *Ramsgate Sands* is returning to Margate. There was even a recent plan to recreate a Victorian bathing machine, with sauna, for visitors to hire, although the Bellman in Lewis Carroll's 'The Hunting of the Snark' might not have approved. As he said of one of the Snark's 'unmistakeable marks':

> *The fourth is its fondness for bathing machines,*
> *Which it constantly carries about,*
> *And believes that they add to the beauty of scenes –*
> *A sentiment open to doubt.*

4

SPIES, FASCISTS AND OTHERS

‡

Coastlines have often been contested ground where the forces of law struggle to maintain their authority. The encounters between customs and excise men and smugglers, for example, is one of the enduring narratives of the East Kent seaboard. Indeed, my earliest awareness of coastal Kent is from childhood books about smugglers on Romney Marsh. The setting was altogether strange, even exotic, there being nothing quite like Romney Marsh on the New Zealand coastline. My misreading of excise men as exercise men made these stories even stranger.

Coastal borders are also places where anxieties cluster, where enemies are believed to lurk. In times of war, or its threat, strangers can be taken for spies. There is a history of writers with heterodox views and, living on the coast, being suspected of conspiring with the enemy. In 1797, with talk of a French invasion fleet off Fishguard, the Home Office sent an agent to spy on Coleridge, Wordsworth and their friends at Stowey, near the Bristol Channel, because they were suspected of signalling to the enemy. The agent described them as 'a mischievous gang of disaffected Englishmen… a Sett of violent Democrats' and thought that Coleridge and Wordsworth's discussions of the philosopher Spinoza referred to a mysterious 'Spy Nozy'. During the First World War, the non-combatant D.H. Lawrence and his German wife, Frieda, were living at Zennor on the North Cornish coast. When two merchant vessels were torpedoed directly off the coast, they were suspected of having signalled to German submarines from the clifftop. Their post was intercepted, their cottage searched and, eventually, they were expelled from the area.

Similar fears were aroused in Thanet in the 1930s as incomers and strangers provoked suspicion of espionage. An article in the *Thanet Gazette* in 1937 headed 'Why are they calling Broadstairs "Little Germany"?' reported that William Joyce, deputy leader of the British Union of Fascists, later the notorious Lord Haw-Haw, had taught at a local convent school and that Hitler's ambassador, Ribbentrop, was a frequent visitor to the town. It was well known that he already rented a holiday bungalow, 'Greenheyes', at nearby Minnis Bay.

Thanet as a haunt of spies and traitors had a famous fictional precursor in John Buchan's novel *The Thirty-Nine Steps* (1915), written while Buchan and his family were visiting relatives holidaying in a house, 'St Cuby', on Broadstairs' North Foreland Estate. Across the road from 'St Cuby' was a flight of steps leading down through the cliff and onto the beach. Buchan began writing fast. Britain declared war on Germany on 4 August 1914; Buchan turned thirty-nine on 26 August; and by November, he'd finished his best-known work.

I'd been following the water's edge around the curving sandy bays of the Thanet headland. At the end of Kingsgate Bay, approaching Broadstairs, I left the beach and climbed the steep coast road up to North Foreland Estate and the lighthouse, which, unlike its counterpart at South Foreland, is still in operation. I wanted to see 'St Cuby' and return to the beach down Buchan's thirty-nine steps.

There seems to be a compulsion to give houses edging the coast a name. Like ships at sea, these houses must proclaim themselves. At North Foreland Estate, as well as 'Seagulls', 'Windy Ridge' and the inevitable 'Casa Mia', there is a taste for Scottish names: 'Inniscarrig', 'Lochiel', 'Currach' and 'Tannochbrae'. Here, at the most easterly point of Thanet, with nothing but wind farms on the horizon, the names are reinforced by flags, the Union flag and that of St George fluttering in the gardens of many of the houses, asserting the identity of the nation on whose clifftop they stand.

North Foreland Estate was established at the beginning of the twentieth century. Although gateless, it is heavily protected, the wide roads and verges and the ponderous sterile villas guarded by the word 'Private', which is blazoned everywhere: private access, private roads, private parking. There's no one in sight but, together with the prominent CCTV cameras, this heightens my uneasy sense of being watched. There are no plaques recording the names of the famous who have lived here either – the privacy of the estate has been defended against history itself – but it was easy to recognise 'Naldera', Lord

Curzon's summer residence after his return from India in 1906, from the photograph in David Seabrook's *All the Devils Are Here* (2003). Curzon had been the last Viceroy of India under Queen Victoria, was briefly Warden of the Cinque Ports (adding Walmer Castle to his portfolio of houses), Foreign Secretary under Lloyd George and altogether a most superior person. 'St Cuby', a more modest holiday house, is just a couple of doors away.

My edition of *The Thirty-Nine Steps* has an introduction by Stella Rimington, former Director General of MI5. She writes that Buchan was convalescing in a nursing home at Broadstairs, that the flight of wooden steps leading down to the beach was later removed and presented to him, and that the novel is 'little more than a hundred pages long'. Buchan and his family weren't in a nursing home; the story of him being presented with the steps is ridiculous – there is a more plausible one that a set of bookends made from several of the steps were presented to Alfred Hitchcock after he had filmed the novel – and the edition Rimington is introducing, a reprint of the original, has 253 pages. It is disconcerting that the sometime head of the organisation responsible for national security could be quite so wrong about so many things. At the very least she might have taken the trouble to check the number of pages.

The top entrance to the steps is surrounded by shrubs, enclosed by heavy steel fencing, and has a sign announcing that access is restricted to residents who are key-holders. The gate, however, was unfastened, and I went down twenty-six steps before meeting a second gate, this one locked. From the bottom entrance, back on the beach, I made my way up ninety steps – some concrete, some brick, some wooden – before hitting the gate that had blocked my descent. A grand total therefore of 116 steps which, if presented to Buchan, would have posed him a problem. On a later visit, the entrance at the top had been locked and, from the beach, another locked gate, twenty-eight steps up, stopped me from going any further. I shook it to see if it would budge and a winking red sensor came to life. The estate is as proprietary about its steps as it is about everything else. Intruders must be kept out.

The Thirty-Nine Steps is concerned with intruders who threaten the nation at a moment when 'all Europe [is] trembling on the edge of earthquake'. Its hero, Richard Hannay, is a much more thorough investigator than Stella Rimington. Towards the end of the novel, the search to identify the sinister 'Black Stone', a German espionage team, and thwart its plan to ring the coastline with mines and submarines has become concentrated on a 'big chalk headland', the Ruff – Buchan's fictional name for North Foreland. The Ruff, a term from cards meaning to trump, is 'a very high-toned sort of place', the epitome of 'the great comfortable, satisfied middle-class world, the folk that live in villas' and who 'like to keep by themselves'.

One of these villas, 'Trafalgar Lodge', is set opposite a flight of thirty-nine steps running down to the beach. Hannay is convinced this is where the Black Stone is based:

> *[A] red-brick villa with a verandah, a tennis lawn behind, and in front the ordinary seaside flower garden full of marguerites and scraggy geraniums. There was a flagstaff from which an enormous Union Jack hung limply in the still air.*

In his first novel *The Power-House*, written in 1913 but published after *The Thirty-Nine Steps*, the hero, Andrew Lumley, speaks of a fear that pervades Buchan's early fiction: 'You think that a wall as solid as the earth separates civilisation from barbarism. I tell you the division is a thread, a sheet of glass. A touch here, a push there, and you bring back the reign of Satan'. *The Thirty-Nine Steps* dramatises this fear.

For most of its 253 pages, Hannay has been on the run from the police, wanted for a murder he hasn't committed. To evade capture, he adopts a bewildering series of disguises, stretching to the limits of credulity the idea that to preserve his identity he must become a succession of others: a milkman, a road-mender, an Australian and so on. By the novel's climax, Hannay's innocence has been established and, restored to himself, he is heading the desperate attempt to save the nation from the Black Stone, whose own virtuosic shape-shifting skills will leave 'men… lying dead in English fields' unless their disguise is exposed and their plans trumped.

Spying on the house from a neighbouring golf course (still there on the slopes between North Foreland and Kingsgate), Hannay is disconcerted by what he sees. Two of the men are playing tennis, for all the world like 'guileless citizens taking their innocuous exercise, and soon about to go indoors to a humdrum dinner, where they would talk of market prices and the last cricket scores and the gossip of their native Surbiton'. They are joined by a third with a bag of golf clubs slung on his back. When Hannay calls at 'Trafalgar Lodge', he finds it full of the things you'd find in 'ten thousand British homes', down to a print of Chiltern winning the St Leger: the house 'was as orthodox as an Anglican church'. At first, he is bewildered by the perfect reproduction of well-bred, middle-class domestic life he finds. But in an uncanny moment of realisation, Hannay sees through the disguises and whistles for the police. Two of the men are arrested, and the third escapes down the thirty-nine steps which he blows up behind him. The German plot is foiled, and the novel concludes: 'Three weeks later, as all the world knows, we went to war'.

Here on the White Cliffs that symbolise the nation, the ease with which the British way of life has been imitated exposes its fragility. The malign German intruders into the nation's front line blend perfectly with the scene, impersonating placid British middle-class existence with casual ease. Only Hannay's persistence and ingenuity can see through the camouflage and put life back into the Union Jack that hangs so limply from the flagstaff of 'Trafalgar Lodge'.

After Curzon's death in 1925, the story of the 'enemy within' on North Foreland Estate moves several doors away to 'Naldera' and gets caught up in Buchan's fictional world. Curzon's daughter, Cimmie, had married Oswald Mosley, then a Conservative MP – 'my sinister son-in-law' as Curzon described him – in 1920. The couple fell out with Curzon and 'Naldera' was left to a stepdaughter who, in 1930, sold it to Arthur Tester, a German with links to German military intelligence.

David Seabrook has written vividly of the chameleon life of Tester, his dodgy business schemes and his network of associates in the British Union of Fascists (BUF). Expensively dressed, sporting a monocle, speaking fluent English though with an unmistakeable German accent, Tester could have been one of Buchan's flashier villains. One of his political and business associates was Joseph Hepburn-Ruston, Audrey Hepburn's father (she herself was born down the coast near Folkestone). Another was Oswald Mosley. Tester claimed to have been Mosley's personal aide-de-camp, and there were local reports of Mosley visiting Tester at 'Naldera'. William Joyce and Ribbentrop were also said to have visited.

Most accounts of the BUF in the 1930s concentrate on its metropolitan activities, and there was certainly nothing like the violent Olympia meeting or the battle of Cable Street in Thanet. But there was a branch of the BUF in Broadstairs, at 23 High Street, and regular parades were organised in the town. According to Seabrook, the members, 'largely... errand boys, complemented by the occasional simpleton', were offered inducements to join: 'a black shirt was, after all, a shirt', he remarks. But Seabrook also cites the case of five young men convicted at Margate Police Court in 1936 for painting 'Jew' and daubing swastikas on a number of local shops.

There was verified German spy activity in Thanet at this time. In 1937, Hermann Goetz was sentenced at the Old Bailey to four years' imprisonment for spying on airfields in Kent. Goetz had rented a bungalow in Broadstairs, near North Foreland Estate. His abrupt departure before the end of his tenancy made his landlady suspicious and she contacted the local police. Special Branch was called in and discovered a camera with film showing pictures of airfields and aircraft. Investigation revealed that Goetz, together with a young woman, Marianne Emig, had been touring the area on a motorbike, sketching the layout and defences of airfields. At his trial, Goetz claimed that he was gathering material for a novel, but he was convicted under the Official Secrets Act. Assuming he was a spy, though clearly an inefficient one, Goetz would have been particularly interested in nearby Manston Airport, which had been a Royal Flying Corps base during the First World War and was to be extensively used in the Second.

By 1938, Tester's political and business activities, particularly his involvement in the European Press Agency Ltd, which was believed to have links with Goebbels, and with Ribbentrop's press attaché, led to questions being raised in the House of Commons. Tester hurriedly left the country, sailing from Southampton to Lisbon aboard his yacht, the *Lucinda*. The *Thanet Gazette* reported that the local police were 'baffled' by his 'mysterious disappearance': 'Local people say that Tester is a spy, and they believe he escaped by rowing out to a submarine before the police could arrest him'. Some of these local people would have been among the hundreds of thousands who had read *The Thirty-Nine Steps* in which, at the end of the novel, one of the enemy agents escapes down the steps and out to a waiting ship.

Tester's story after departing 'Naldera' is as incredible as anything in a Buchan novel. The *Lucinda* was seized by the Royal Navy at Naples while Tester was visiting Mussolini in Rome. He then settled in Romania where he worked for the Abwehr in Bucharest, specialising in interrogation. Late in 1944, he was shot dead in Transylvania by a border guard while trying to escape into Hungary. The *Thanet Gazette* reported he was carrying a passport signed by Hitler. Soon there were rumours that he'd survived, faking his death by substituting a body dressed in his clothes and with his watch and cigarette case. This was pure Buchan, and the authorities took pains to still the rumours. At Britain's request, the Russians exhumed the body and an unlikely sounding

Ramsgate dentist, Brigadier J. Marley Stebbings, examined Tester's teeth and confirmed his identity. Stories of Tester's afterlife persisted nevertheless, and in the 1950s, he was reported to be gun-running in the Middle East.

Tester had set Thanet abuzz. As the *Gazette* reported when confirming Tester's death: 'It is believed that Dr Tester was a Gestapo chief in Kent preparing the ground for the control of the country in the event of a successful invasion'. *The Thirty-Nine Steps* had provided the script for a real-life spy story in the very spot where the climax of the novel is set. Perhaps Tester had even read the novel. The tunnel of steps leading from North Foreland Estate to the beach below would have made Curzon's house a tempting purchase. 'Naldera' remained empty during the war, was damaged by a bomb that hit North Foreland, looted and put up for auction at the end of the war. An oil painting by Turner was said to be among the contents of the house but this, like much else in Tester's life, proved to be false. Since then, the excitement of life at North Foreland Estate has subsided.

North Foreland and Sandwich Bay Estates, like the towers of Reculver and Richborough, and the lighthouses of North and South Foreland, offer a pair of matching features around this coastline. Sandwich Bay Estate is also a bleak and unlovely place. Dead flat, exposed, acarpous, it resembles somewhere the undead might go for their summer holidays, a ghostly place picketed by admonitory notices – 'No camping', 'No overnight parking', 'No fires', 'No L Drivers', 'No entry' – protected by CCTV and by its uncanny feel. Just to walk around it seems like an offence. The large houses of the estate – 'The Dunes', 'The Lodge', 'Fifth Tee' (this one backs onto Royal St George's) and 'Top Edge' (the home of a cricket enthusiast) – resemble mausoleums in a vast garden cemetery but a cemetery from which flowers have been banished. Sandwich Bay itself is noted for its wild flowers: the marsh and lizard orchids that grow along the wide verges of the estate and the yellow sea poppy, white campion, mallow, wild fennel and yarrow that flourish in its dunes. The gardens of the houses, by contrast, are strangely unplanted, with little to relieve the wide acres of severely mown lawns.

Like North Foreland Estate, the most formidable-looking residences hog the seafront and have a history of association with pro-German sentiment in the years before the Second World War. *Pares inter pares* is 'Rest Harrow', built in 1910 for Nancy and Waldorf Astor. Designed by a pupil of Lutyens, this fourteen-bedroom Arts and Crafts style house (Nancy Astor called it her 'seaside cottage') remained in the Astor family until a few years ago. During the

1930s, one of the Astors' neighbours was Captain Robert Gordon-Canning, a close friend of Oswald Mosley and the owner of a fascist newspaper, *Action*. On their other side was Lord St Just, son of a former governor of the Bank of England and Conservative MP for the City of London. Others nearby included a member of the Spencer-Churchill family and the future Conservative prime minister Harold Macmillan and his wife Dorothy. Drawn there by its proximity to two exclusive golf courses – the Royal St. George's and the Prince's – Sandwich Bay Estate has been described by Clare Ungerson in *Four Thousand Lives: The Rescue of German Jewish Men to Britain, 1939* (2014) as 'Mayfair by Sea'.

Nancy Astor was a mass of contradictions. The first woman to take a seat at Westminster, a Tory MP from 1919 to 1945, a Christian scientist; she disliked Catholics, Jews, communists, the French, alcohol and music. She described herself as a feminist, tried to found a women's political party, campaigned for child welfare and the education of the poor, set up nurseries in her Plymouth constituency and helped establish a training college for nursery school teachers. She was a keen squash player – 'Rest Harrow' had its own court – and golfer. When visiting the Inner Hebrides island of Jura, I was shown a flat-roofed building on the Astor estate in the north of the island from where, I was told, she would drive golf balls into the sea. Astor believed in the health-enhancing effects of seawater and had it piped directly into the two grand bathrooms attached to the master bedroom at 'Rest Harrow'. She was a society beauty with a wide circle of male friends but a declared dislike of sex. Curzon was one of a number who fell, unrequitedly, for her.

Astor was at the centre of German appeasement circles during the 1930s and her house parties at Cliveden, the lavish Buckinghamshire estate that had been a wedding present from her father-in-law, were notorious for their alleged support of Hitler. She certainly knew Ribbentrop well. In March 1936, just after the German invasion of the Rhineland, she threw a party at her London residence in St James's Square. She had Ribbentrop placed next to her at dinner and later in the evening persuaded him to take part in a game of musical chairs. Astor whispered to her guests that they 'must let the Germans win'. Shortly after this, Ribbentrop met leading pro-appeasement figures at 'Rest Harrow', among them the minister for co-ordination of defence, Sir Thomas Inskip, and Tom Jones, former Deputy Secretary to the Cabinet and a confidant of the prime minister, Baldwin. Ribbentrop was trying to arrange a face-to-face meeting between Baldwin and Hitler. The discussions at Sandwich Bay lasted until 1am and the participants decided that the map of Europe might eventually be redrawn by agreement between Britain and Germany. A memorandum to this effect was sent to the foreign secretary, Anthony Eden.

There were many reasons for pro-German feeling in the mid-1930s: anti-communism (the director general of MI5 at this time explained that fascism was 'a natural reaction from communism'), the belief that Germany had been harshly treated at Versailles, a desire to avoid another war but also admiration for the order and discipline that Hitler was thought to have brought to the country, reflex anti-Semitism, deliberate blindness to the brutality and persecution of the Nazi regime and an offhand disdain for democratic opinion. The most subtle analysis of the world of appeasement that I know is Kazuo Ishiguro's novel *The Remains of the Day* (1989). I remember the scorn poured on it by Germaine Greer – 'a novel about a *butler*', she remarked disparagingly – and its political acumen continues to be little remarked on by the many commentators who have admired its psychological realism, its comedy of manners, its narrative voice and so on. Stevens, butler to Lord Darlington, a leading appeaser, witnesses the manner in which 'the great decisions of the world' are arrived at 'in the privacy and calm of the great houses of this country'. Among the guests at Darlington Hall who convene to discuss the European issues of the day are Lady Astor and George Bernard Shaw, a close friend of hers and a regular visitor to 'Rest Harrow'. Other real-life guests at Darlington Hall include the foreign secretary, Lord Halifax, and Ribbentrop – 'no stranger to Darlington Hall'. Neither was Oswald Mosley, although Stevens defensively describes him as having visited only in the early days of the British Union of Fascists (BUF) before 'it had betrayed its true nature'.

The Remains of the Day, like that late-night meeting at 'Rest Harrow', captures how sections of the British upper classes flirted with Nazi Germany and cooperated with Ribbentrop in his attempt to secure an Anglo-German alliance. Stevens' retrospective touchiness about Mosley's presence at Darlington Hall illustrates the blurred lines between advocates of appeasement and the leaders of the BUF in the 1930s. Nancy Astor's neighbour, Gordon-Canning, old Etonian, Oxford graduate, late of the 10th Royal Hussars, was impeccably qualified to be part of Mayfair by Sea. He was also Ribbentrop's golfing partner, best man at Mosley's second wedding (to Diana Guinness, nee Mitford) in Berlin where he met Hitler, as well as accompanying Mosley to Rome to negotiate a subsidy to the BUF from Mussolini. Canning was interned throughout the war and on his release successfully bid for a bust of Hitler when the contents of the German Embassy were auctioned.

Another leading member of the BUF lived in Sandwich itself, just inland from Sandwich Bay Estate. Lady Grace Pearson was from an aristocratic family and her brother, Henry Page Croft, was a very right-wing Conservative MP. She lived in one of Sandwich's most gracious homes, 'Manwood Court', and owned other properties in the town, one of them home to the local offices of the BUF. She stood as a parliamentary candidate for the BUF in Canterbury and knew Mosley well. A local resident, whose father was the butler at 'Manwood Court', told Clare Ungerson that Lady Pearson and Mosley had been lovers.

Fascism in Sandwich involved its civic leaders too, foremost among them George Solley, farmer, magistrate and several times Mayor of Sandwich. Mosley and the BUF had attracted the support of local farmers because of their opposition to imports of cheap food and their demand for lower taxes. Solley co-founded the Sandwich branch of the BUF with Lady Pearson and also worked with Canning. When William Joyce came to speak in the town, he chaired a packed meeting in the Sandwich Guildhall.

By the mid-1930s the Board of Deputies of British Jews' defence committee, which kept a close eye on the BUF and received reports on its activities from across the country, mainly from large metropolitan areas, was also receiving them from Thanet and Sandwich. These became more frequent and detailed after a transit camp for Jewish refugees was established just outside Sandwich in 1939.

Until the end of 1938, most German Jews escaping Hitler either had the means to support themselves or were individually sponsored, but after *Kristallnacht* in November of that year, there were many without resources or sponsors in urgent need of a place of refuge. After intensive lobbying by the Central British Fund for German Jewry (CBF), the government agreed to accept groups of refugees on condition that responsibility for them lay entirely with the CBF. It also insisted that permanent residence wasn't on offer – they were to be regarded as in transit – and that only in exceptional circumstances would they be permitted to work.

One of the CBF executives had designed the Navy, Army and Air Force Institutes (NAAFI) dining hall at the Kitchener Camp at Richborough during the First World War and remembered the site. The camp, although dilapidated, was empty and available, and its proximity to Dover where the refugees would be landing made it an ideal location. It was quickly made habitable and, within a few months, around 3,500 Jewish refugees had arrived there. They were entirely men, most of them young – a demographic felt to be best suited for a transit camp.

Sandwich in 1939 had a population of just under four thousand. The Kitchener Camp therefore almost doubled the size of this quiet and economically depressed town. The CBF was aware of the hostility that such a concentration of Jews might provoke, especially given the presence of the BUF and the age and gender of the arrivals. But reports from Sandwich were, on the whole, reassuring. The local Chamber of Commerce thought the camp was likely to be good for business. Lady Pearson, the CBF was told, was disliked among the working people of the town. The BUF headquarters and bookshop in Strand Street was 'devoid of stock' and seldom with anyone in attendance. The only member of the BUF encountered by the CBF reporter was Lady Pearson's gardener.

But there was some opposition at first. A demonstration of flag-bearing Blackshirts from Sandwich and Deal marched to the camp. Anti-Semitic correspondence began to appear in the *East Kent Mercury* focused on the dangers of interbreeding between Jew and Gentile. One letter warned of mental retardation, stammering and asthma as likely consequences of such 'miscegenation'. When the editor refused to publish any more letters of this kind, Canning continued the attack in his own weekly, *Action*. Under a banner heading 'REFUGEES', he described the disruption to the life of Sandwich caused 'by the intrusion of this Central European mass into its quiet and orderly existence' and how 'the presence of this foreign excrescence' was endangering 'the good name of this ancient town [and] its repute as a beauty spot of England'. Canning also invented a story of a young local girl having been assaulted by two of these 'aliens': 'Sex-starved they may be, but it is not for British womanhood to appease their appetites'. And in a rhetorical move familiar again today, he wrote of the need to protect the welfare of the town and particularly the needs of its unemployed.

By and large, however, the arrival of these strangers was welcomed, not least because they livened up a rather dreary little town. Although the purchasing power of the arrivals was limited, local shops did enjoy new custom. But it was sport and music that did most to bring town and camp together. Men from Kitchener joined the local string orchestra and Sandwich, Margate and Ramsgate football clubs; there was friendly table tennis rivalry between town and camp; joint cycling activities were arranged. Concerts at Kitchener attracted as many as a thousand people from the town. And the Kitchener Camp jazz band played at fetes and dances up and down the coast. Being prohibited from paid employment meant the band was free and it became very popular, although local musicians did complain they were losing out on gigs. The refugees were also invited to the Betteshanger Colliery Sport and Social Club after which the miners sent money to the Kitchener Camp Welfare Fund. This mingling fell below the gaze of the smart set, but Nancy Astor visited the camp and took out a subscription to its newspaper. By now, Anglo-German friendship circles were falling quiet and it was only overt anti-Semites and fascists like Canning and his followers who publicly objected to the refugees.

Kitchener Camp went as quickly as it had come. At the beginning of 1940, it became a mixed military and refugee camp and around half of the refugees enlisted in the British Army. Others waited for emigration to the United States or Palestine or were reluctant to sign up for fear of what might happen to their families back in Germany. The fall of France in May and June 1940 put the Kent coast directly in the front line and Sandwich was now seen as too vulnerable for large numbers of aliens, no matter how friendly, to be quartered there. The enlisted men were moved to the south-west of the country and the remainder were sent to an internment camp on the Isle of Man – a radical change of status and treatment. Kitchener had been a place of transit, not internment.

The sudden emptying of the camp was a shock for the residents of Sandwich. They had become used to the presence of the Kitchener men and the social attachments that had formed were brought to an abrupt end. Several of the men returned to Kent at the end of the war, and one, Philip Franks, settled in Deal where he married and worked at Betteshanger Colliery. Otherwise, however, the Kitchener Camp faded from sight and memory. Some of the buildings became industrial premises and, in 1953, the land around the camp was engrossed by the pharmaceutical company Pfizer. The Association of Jewish Refugees installed a memorial plaque on the inner arch of the Sandwich Barbican, which curiously neglected to mention that the inhabitants of Kitchener Camp were Jews.

The short-lived and, until Clare Ungerson's book *Four Thousand Lives*, almost forgotten history of the transit camp is more than just a footnote to the story of Britain's treatment of refugees in the last hundred years. The prevalence of anti-Semitism in British society at large worried everyone involved in establishing the camp and the presence of the BUF in the towns of the East Kent coast justified their anxiety. But in the event, most of these fears proved groundless. A small, conservative town with marked class differences not only tolerated but in the main welcomed the arrival of strangers with a different culture and from a nation the people of Sandwich had been at war with only twenty-five years earlier.

It might seem facile to contrast the acceptance of these 'friendly aliens' with the current suspicion and resentment of refugees in coastal areas such as East Kent. There were circumstances then that seem not to operate today. For example, Britain was at war with the country from which they had fled, but the same could be said of refugees from several Middle Eastern countries now. Kitchener men were not allowed to work and so Canning's faux concern for the threat they posed to 'the needs of the unemployed' had no purchase, but such is the case for asylum seekers today. Of course, there are real differences. The social background of the Kitchener men was mainly lower middle class and artisan – shopkeepers, traders, tailors, skilled workers – and this roughly matched the social make-up of most of Sandwich's inhabitants. You can see this reflected in the shared sports and entertainments enjoyed by the two communities.

Yet there is something exemplary in the story of how Sandwich accepted the refugees. In 1941, George Orwell challenged the nation: 'England has got to be true to herself while the refugees who have sought our shores are penned up in concentration camps, and company directors work out subtle schemes to dodge their Excess Profit Tax'. A couple of years earlier, the people of Sandwich had ignored the ugly politics of their coastline and welcomed several thousand Jewish refugees from Hitler's Germany into their town. Borders can be places behind which people find sanctuary as well as barriers to keep others out.

5

SCRAP LANDS

‡

Waste, waste, everywhere. Unlike the mounds of plastic near the Stour Estuary, this is degradable or convertible or refurbishable or renewable stuff, brought to the scrap lands between Richborough and Sandwich to be alchemised. There's money in waste, as Dickens' Golden Dustman, Noddy Boffin, understood. But there are no scavengers picking their way through these mounds. The businesses lining both sides of the road are fenced and guarded.

I'm on the A256 between Richborough and Sandwich. It's a dead-straight dual carriageway, a bit of a racetrack. There's a wide footpath both sides of the road and a cycle lane for those who can tolerate the roar and fumes of fast-moving vehicles. The massive remains of Richborough Roman Fort lie just inland and the skeletal remains of Richborough Port are hidden away behind the businesses that cut me off from the coast.

Ambrosetti UK Ltd is making it new, refurbishing dilapidated Fords, rows of vehicles shiny and once again saleable coming off the reproduction line. Next door is a vast enclosure like an airport car park but filled with hundreds of container lorries – Tijsen Transport, G. Mariani & Co, Altreks – detained behind high fences. The security guards at the entrance to Ambrosetti tell me it's a holding station for vehicles that have been seized by Customs. Many of the lorries look as if they've been here for a long time. Weeds are growing up around them. This is the transport equivalent of a detention centre.

Timberlake Recycling Merchants and Waste Paper Depot is a rusting corrugated iron building with gaping holes in its sides and roof, in need of recycling itself, looking as if a tornado has gone through it. Its yard is dense with enormous bales of compressed waste paper, like huge chunks of fallen chalk, as large as the heavy machinery that pushes them around. Next door is Stonar Cut, a short channel connecting the two arms of the Stour and where the old Richborough Port began. There's a large, rusting barge sunk in the mud of the Cut, as if it too is waiting to be salvaged or recycled. Across the sluice there's a sign reading 'Thanet Skips'. That isn't how Thanet had struck me.

On towards the old Pfizer complex, now reconfigured as Discovery Park, massive concrete tanks give off a thick smell of effluent. A snail is baking to death on the footpath; a solitary woodlouse takes cover under the scanty grass verge. The yammer of the traffic and the heat coming off the footpath is getting into my head, under my skin. I turn off into Discovery Park, 'The Hub for Science, Technology, Business and Enterprise'. The message, for those who don't get it the first time, is underscored by buildings with names like 'Innovation House'.

It's lunchtime and lean, fit-looking employees are jogging purposefully around the complex. I'm a long way from Cliftonville. The site is being upgraded and adapted, made over to accommodate many small enterprises rather than a single large one (Pfizer had been the biggest employer in East Kent). Acres and acres of parked cars are glinting in the harsh light, another airport car park but no weeds here. Through Discovery Park and out the other side is a short sad trail of lost little businesses – a remaindered small press, a workshop that once made garage doors, an abandoned salvage tip – and a couple of densely overgrown buildings, remains of the Kitchener Camp. I get a glimpse of Stonar Lake through the trees but there's no way in behind the heavily locked gate and high barbed wire fence. The lake is for private fishing.

Sandwich figures in Richard Aldington's bitter brilliant novel of the First World War, *Death of a Hero* (1929), as the 'commonplace little money-grubbing town' of Hamborough, where the fourteenth-century Barbican is regarded by the inhabitants as an obstacle to progress, blocking a proposed new road, and saved only because the thickness of its walls make it too expensive to demolish. But it has survived, and so has the rest of this preserved-in-aspic town. Crossing the Stour and entering the town through the Barbican and past the memorial plaque to the Kitchener Camp, tidal and flood protection work is in full swing. Just along the quay is the medieval Fisher Gate. No discovery going on here, no alchemy; Sandwich is dedicated to preservation and the past.

I stroll through the town, the Norman tower of St Clement's jutting above the houses to one side, that of St Peter's to the other. Elizabeth I gave permission to twenty-five Flemish families, Protestant refugees, to settle in Sandwich, and they made the town a centre of the weaving industry. These incomers built what is now the United Reformed Church in the cattle market, and when the tower of St Peter's collapsed in 1661, they also rebuilt that. St Peter's then became their place of worship, known as the 'Strangers Church'. Ford Madox Ford described Sandwich as 'a Low Country seventeenth-century town', a smaller version of Bruges, and there is Flemish influence all around me in the houses.

I'm heading for Tom Paine's cottage in New Street. Paine lived in Sandwich at the end of the 1750s. He moved here after a brief stint in Dover and set up business as a stay-maker – producing ribbing for women's corsets – which had been his father's trade. He was married in the 'Strangers Church' to a local woman, Mary Lambert, who worked as a servant in the household of a prominent draper and former mayor of Sandwich, Richard Solly – probably a forebear of the Sandwich fascist, George Solley. Paine's business failed and he and Mary moved further around the coast to Margate where she died in childbirth, the infant with her. 20 New Street is a small, low-slung, Grade-II-listed seventeenth-century cottage with a succinctly worded plaque: 'Tom Paine's cottage c. 1759. Author of "Rights of Man". Inspired the American Declaration of Independence'. You can rent it for self-catering holidays.

Sandwich is an unlikely place to be linked to something as epoch-making as the American Declaration of Independence. Paine challenged most things that Sandwich stood for then and now. A radical democrat and republican notorious for his attacks on established religion, a significant figure in the French as well as the American revolution, memorialised in a town that looks so steadily to the past. Yet Sandwich has a tradition of accepting incomers – Flemish weavers in the sixteenth century, Jewish refugees in the twentieth – as well as experiencing the coming and going of traders from around the coast of Britain and across the Channel and the North Sea. Paine, an incessant traveller and migrant who never owned a property until the last years of his life, a transient who always followed the politics rather than the money, isn't entirely out of place here.

In fact, Paine still doesn't have a settled place of rest. William Cobbett, some years before he came this way during the journey that resulted in *Rural Rides* (1830), exhumed Paine's body which had been buried at New Rochelle in New York State and brought it back to England. After Cobbett's death, the remains were dispersed – the skull to Australia, the brain back to New Rochelle, most of the skeleton secretly buried by a Unitarian minister somewhere in the Manchester area, or so the story goes – a strange example of the migration and recycling of body parts. As Thomas Browne remarked in *Urne-Buriall* (1658): 'But who knows the fate of his bones, or how often he is to be buried'.

Back up the other side of the A256 towards Ebbsfleet where I've left my car. Copart is yet another airport car park, but one that's been droned or strafed, a wasteland of salvaged vehicles, the biggest car crash ever. 'We Can Buy Your Car! Any Vehicle; Any Value; Any Mileage; Any Condition', it announces. Oddly, the gates carry a sign in Polish, '*Nie Parkowac*'. Are Poles especially likely to block the entrance? There are other motor graveyards and resurrection sites along this strip of coast. 'Zen Motor Factors', just inland from here, sounds as if it's dedicated to traffic calming. Peering in through the high wire fence, Copart is like an A&E ward for traffic after a major disaster: polythene sheets over smashed windscreens, dented bodywork, vehicles ailing and sunk on their axles. Even as I stand by the entrance, lorry-loads of written-off cars are coming in through the gates past the '*Nie Parkowac*' sign. I imagine the land along this strip of coastline steadily filling with discarded vehicles until Copart nudges up against Richborough Roman Fort.

Copart's founder, Willis Johnson, is described on the company website as having lived the American Dream and, through Copart, enabled others to live it as well. His book, *Junk to Gold*, describes how the business grew from a single junkyard in Vallejo, California to a $4.5 billion 'international high-tech inspired auction company'. Johnson explains his success as 'partly due to the lessons my father taught me and partly due to God's hand guiding me along the way'. I remember Jonathan Aitken's Damascene experience on the nearby Sandwich shoreline just a mile or so away. The hand of God has been busy round here.

Beyond the mausoleum of Copart, there's lots more waste: 'Household Waste Recycling Centre: "Recycle for Kent"'; 'TW Services ("Let's Sort It Out")' promises to 'reclaim, recycle, reduce, reuse'. Amid all this, the natural world is doing its own recycling. Wild flowers – daisies, clover, buttercups, dandelions, a few early poppies – are sprouting along the edge of the A256. Then, inexplicably, there's a small insecurely fenced-off area in the verge with a different kind of sign from those I'd been noting: 'Keep Out – Protected Flowers'. A clutch of pretty white flowers, like snowflakes, are sheltering behind the fragile fence. I've no idea what they are or who is trying to protect them. The notice is unsigned. Sitting up behind 'Stevens & Carlotti Metal Fabrication', I see my old companion, the wind turbine.

Closer towards Thanet, the recycling of solid waste modulates into the conversion of heavenly light into power. 'Solar PV Systems' is a prelude to the solar farms that are mushrooming all over Thanet. This isle, as Turner knew, is full of light. Caliban, in his tender speech, 'Be not afeard', recalls a dream in which 'The clouds methought would open, and show riches / Ready to drop upon me', and so it is proving. Back at Stonelees roundabout, near where I'd started my walk, there's a farm of these smoky-grey, gently angled panels, like the flaps on the wings of an aircraft, spreading across the land in orderly lines, hugging and shading the ground. Another solar farm runs up the slope towards Thanet's ridge like the smooth face of a slate quarry. The sun is still glaring; it's fertile weather for solar power. There are solar farms everywhere across Thanet, some large-scale commercial enterprises, others private and small-scale, like greenhouses in a back garden. In *Rural Rides*, William Cobbett's description of reaching Thanet after walking from Dover and through Sandwich is the only bright moment in an otherwise rancid account of poverty and corruption: 'When I got upon the corn land in the Isle of Thanet, I got into a garden indeed. There is hardly any fallow... It is a country of corn'. It's a long time since Thanet was a garden, but the harvesting of wind and light are making it so again.

6

CHALK

‡

The white stuff that forms and defines much of the coastline around which I am walking seems inseparable from martial images of defence. Robert Macfarlane, introducing the republication of Jacquetta Hawkes' *A Land* (1951), describes how she ends her book at the chalk cliffs of the Channel coast: 'Britain's Cretaceous bastion, its white shield raised against invaders', although this isn't really how she describes them at all. Carol Ann Duffy's poem, 'White Cliffs', commissioned by the National Trust to celebrate its purchase in 2012 of the stretch of coastline either side of South Foreland Lighthouse, reaches for a similar figure of speech in its opening lines: 'Worth their salt, England's white cliffs / a glittering breastplate'.

Chalk, though, is a deceptive substance, apparently solid but full of faults, inherently flawed and readily infiltrated by water, which seeks out its cracks and joints to form fissures, shafts, sinkholes, conduits and caves. Auden was much closer to its essence in 'In Praise of Limestone' when he wrote: 'When I try to imagine a faultless love / Or the life to come, what I hear is the murmur / Of underground streams, what I see is a limestone landscape'. In the national imagination, the White Cliffs stand for strength, endurance, permanence; they underwrite the nation's island story and guarantee its historical continuity. In situ, however, they are porous, fragile, mutable, eroding and forever being undermined.

In geological terms, the White Cliffs are still youthful. They began to form during the Cretaceous period, between 140 and seventy million years ago, when sea levels rose, leaving only the Welsh and Scottish Highlands uncovered. Chalk formed as microscopic marine plants extracted calcium carbonate from the sea and their tiny skeletons – coccoliths – accumulated on the seabed. As the sea level remained constant, chalk mounted layer by layer, increasing at the rate of one foot in thirty thousand years. Jacquetta Hawkes described the process this way:

> *If, enjoying the sun, a child leans against the cliff at Folkestone, his small figure will span the accumulation of one hundred and twenty thousand years. And yet, knowing this, still my imagination will so speed up the process that I see it as a marine snowstorm, the falling of flakes through one of the clearest seas ever known.*

Chalk is such a variable substance, its face changing almost every step of my way around the Thanet coast. At Botany Bay, huge chalk stacks have become detached from the main part of the cliff, standing apart like the towers of Reculver. A little further on towards Broadstairs, at the north end of Kingsgate Bay, a natural arch has formed in the steep chalk cliff, an elegant portal through which the walker passes. At the north end of Louisa Bay there are several recent chalk-falls, the fresh white gashes in the cliff looking vulnerable to further collapse. On towards Ramsgate, the cliff becomes a great white rampart, solid and indestructible enough to warm the heart of the most fervent Little Englander.

The chalk cliffs I know best and feel closest to run from Kingsdown at the south end of the Deal shoreline on past St Margaret's Bay to Dover. Richard Aldington evokes this landscape, 'the edge of one of the long chalk downs of England', in *Death of a Hero*:

> *The chalk was ridged in long parallels, like the swell of some gigantic ocean arrested in rock. The ridges became more abrupt and violent near the coast, and ended in a long irregular wall of silvery-grey chalk, poised like a huge wave of rock-foam for ever motionless and for ever silent.*

In Aldington's sublime description, the ocean-swell of the land forms the crest of a wave that will never break. The point of view is from the edge of the cliff looking out across the Channel, but the description works just as effectively if imagined from below, looking up.

I'd begun walking the water's edge from Kingsdown to St Margaret's. The margin is narrow, and the walk can only be made at low tide. The cliff –

rugged and fragile, hard and soft – looms above and bends over me. It forms gnarled and sagging faces, admonitory and forbidding, oddly like those that can sometimes be seen in ancient trees. Looking up at the thin white line of the cliff edge and the calm blue sky above, I feel as if I'm upended and falling backwards. It's like having vertigo from below. The sound of breaking waves and the outgoing tide is caught and amplified by the pits and hollows of the chalk cliff, producing unusual acoustic effects. Chalk is a natural recording studio with its own sound world.

Clambering over the chalk field takes most of my attention. Head down, watching my footing, the chalk left glazed and slippery by the withdrawing tide, walking the shore becomes as mesmerising as trudging over endless snow. Until, that is, I reach Coney Burrow Point where an enormous fall has brought down great chunks of the cliff-face spilling out into the Channel to form a new headland, slightly increasing the fractal distance of the coastline. I scramble up the newly formed spur, worried about more falls (is there any warning before a slippage? The sound of cracking? A rumble?) and aware that if I slip and sprain my ankle, I'm stuck. I've seen no one since beginning the walk and soon the tide will be turning. On top of the saddle, a sturdy thorn has taken root. It must have come down with the cliff fall. There are more traffic cones on the shoreline below me, reminding me of those I'd seen at Sandwich Bay. Where are they all coming from?

It becomes like hillwalking where the top of each ridge exposes a yet higher one. Close to St Margaret's Bay, just a couple of hundred yards before the end house that Noël Coward and later Ian Fleming owned, is another and bigger cliff fall to climb. Looking up at the massive scoop left by the slippage, I know exactly where it's come from. The clifftop path between the Dover Patrol Monument and the steps that drop down into St Margaret's Bay had until recently a seat where one could rest and look down onto the shingle beach, The Pines Garden, and up towards South Foreland Lighthouse. The Pines Garden Museum has a very competent painting of Coward's that captures precisely this view. During the rains of last winter, this seat had vanished over the edge. I look around for it, wondering if I might be able to rest on it at the bottom as well as the top of the cliff.

Looking up at the cliff, I feel precipitously inside it, lost in a white-out. One of Hopkins' 'terrible sonnets' echoes in my head: 'O the mind, mind has mountains: cliffs of fall / Frightful, sheer, no-man-fathomed… / Nor does long our small / Durance deal with that steep or deep'. An afternoon in the shadow of the cliffs, clambering over the evidence of their instability, keeping an anxious eye on when the tide would turn, has exhausted my durance of the steep and deep of this coastline. I've been awed by the massive solidity, the monumentality of the chalk cliffs, while worrying all the time that they might fall on top of me.

These cliff falls that blocked my way are contradictory. On the one hand they're eroding the coastline, nibbling away at the border of the nation. On the other, though, these slippages form new hoes jutting out into the Channel and extending the nation's margin. The more jagged the coastline becomes, the greater its fractal distance. It's as if the process described by Jacquetta Hawkes – one foot in thirty thousand years – is accelerating, but inwards and outwards rather than upwards. The White Cliffs are metamorphic, shape-shifting, not an immutable white shield or glittering breastplate.

One of UKIP's billboards for the European elections in 2014 had a photograph of Nigel Farage and a phalanx of his supporters at St Margaret's Bay massed in front of the White Cliffs, up and through which ran an escalator. This call to lower the portcullis and man the ramparts, using the White Cliffs as a synecdoche of a nation that has dropped its defences, is a travesty. These chalk cliffs are an archive of many different stories, of invasion and defence certainly, but also of the constant flow of people inwards and outwards that defines this coastline, and of the sanctuary frequently offered to strangers. The White Cliffs need to be reclaimed. The beauty of their shimmer from a distance, their unstable face from up close, their mutability and impurity – flints and fossils, thorns and traffic cones – are a more variegated and inclusive image of the nation than any currently on offer.

Jacquetta Hawkes offers a very different vision of these cliffs on the last page of *A Land*. In closing her account of how the 'pieces whose shaping in time by geological processes, by organic life, by human activity and imagination' have made Britain, she turns to 'the long line of the chalk cliffs':

> *Into them I must set esplanades and bungalows, hotels and boarding houses; fishing towns and villages; docks, jetties and piers; estuaries thronged with pleasure craft, and crowded ports, and round them all the movements of the small craft, the coming and going of great ships. So I have tried to celebrate the creation of this land and our consciousness of it…*

No mention here of ramparts, white shields or glittering breastplates, nothing martial, but instead the picture of an abundant, multifaceted, ever-changing and peopled coastline, set in the long perspective of geological time which Hawkes renders so brilliantly:

> *While I have written the sea has swallowed a gobbet of land in one place, released a few square yards in another; there have been losses and gains in the flow of consciousness. Again I see the present moment as a rose or a cup held up on the stem of all that is past.*

UNDERGROUND

7

FAN BAY

‡

I had been intrigued by the National Trust excavations I'd stumbled across at Fan Bay and remembered that volunteers were needed. I contacted Jon Barker, a National Trust manager at South Foreland Lighthouse, who told me that in the late 1970s Dover Council had buried and backfilled the entrance to the underground shelter and entombed the sound mirrors in hundreds of tonnes of spoil topped with chalk, part of a designated 'eyesore clearance programme'. History had been redacted and the past interred. Jon's team were now uncovering that past and he said I'd be welcome to join them. Like Alice lured by the white rabbit, I was led underground.

There were twenty or thirty volunteers on my first morning. Everyone else seemed to have been there before: a knot of men from the Kent Underground Research Group; a middle-aged woman wearing a Wealden Cave and Mine Society T-shirt; most with their own hard hats and overalls. These people were serious about holes in the ground. I was the new boy. There was a warm autumn sun over Fan Bay but heavy mist out in the Channel, as if we were sheltered behind a white curtain draped from the sky.

As a new volunteer, I was taken into the tunnels, dropping down a steep flight of steps that seemed bottomless, like descending a darkened never-ending tube escalator. Slowly, I began to make out this underground world. The walls and the steps had survived everything Dover Council could throw at them, and the carefully arched roof was several feet clear of my head – a relief because I'm tall and had worried about feeling claustrophobic or banging my head. I'd been given a torch and a hard hat with a light, and temporary lighting was rigged up at points along the tunnels. I could see what was immediately around and in front of me but had no sense of where I was going and the tunnels closed up behind as I moved along them in an ampoule of light. I was to become familiar with, even to enjoy, the disorientation of going underground, but on this first morning, I felt engulfed and lost. There was nothing to navigate by, no thread as to where I was. I glimpsed smaller tunnels branching off to the side and some old, unfinished diggings as well. At least I hoped they were diggings and not chalk-falls.

Sam, my guide, had been with the volunteers since the beginning and kept up a stream of information as we made our way through the mazy tunnels. The shelter had been dug and the accommodation completed in three months in 1940–41, used intensively for the duration of the war, then sealed off for the next seventy years. I felt like a tomb-raider. 185 soldiers and four officers had lived and slept down here. Many of the 172 Tunnelling Corps, Royal Engineers who had dug the tunnels, were miners and they lined the walls with colliery arch and metal sheeting that will last for centuries. The heavy locked door we eventually came to at the far end of one of the tunnels had once opened out into the steep hillside where the sound mirrors were still buried.

The deep shelter reminded me of Henry Moore's drawings of Londoners sheltering in the underground during the Blitz. His *Shelter Scene: Bunks and Sleepers* (1941), with its three layers of bunks, was exactly as the Fan Bay accommodation had been arranged. In another drawing from this time, *Tube Shelter Perspective*, the sheltering figures are spread out on the tube platform like corpses after an atrocity, pale and ghostly as the Morlocks in H.G. Wells' *The Time Machine* (1895) – denizens of the underworld. The ceilings were higher and the tunnels wider in Moore's drawings, but their narrowing perspective and cylindrical contour had the same enclosed feel as the Fan Bay shelter.

Most of the rubble had by now been cleared from the tunnels – thirty tonnes of spoil removed by hand – but the last of it was still there, stashed in bags waiting to be taken up the steps. No one comes up from below empty-handed. Wheelbarrows left down here were needed on top today, and so Sam and I each took one and hauled it back up to the surface – my first lung-bursting experience of a climb of 125 steps I came to know well.

While we'd been underground, the curtain of mist over the Channel had opened and the sky-reflecting sea had become the colour of sapphire. Cap Gris-Nez, with its matching white cliffs, stood out clear and bright. The area around the entrance was an archaeological site. Several teams were clearing away Dover Council's dumpings to uncover the original brick surrounds of the entrance. I joined a group trying to unearth the plant room which had provided ventilation for the tunnels and which an old map of the site showed had been to one side of the tunnel entrance. For several hours we found only spoil and chalk. Then the first signs of brick – an exciting moment. We dug and scraped furiously like dogs sniffing out a bone. A hole opened up around the brickwork, a wide cavity down which we could see the exploded remains of the plant room. Its solid concrete roof had been blown in; thick, tangled, steel reinforcing rods were mixed together with smashed lumps of concrete and many tonnes of spoil. Removing all this would be work for our mechanical digger.

By the afternoon I'd begun to feel part of a team, a small, dedicated community. Jon oversaw the work but with a light hand, and everyone seemed instinctively to know what to do. There was an informal shift system that seemed naturally to happen. You'd be banging away with a pick for a while and then someone would come to take over. I fell in with the rhythm of this, relieving someone with a barrow who'd been taking away the spoil. And so it persisted all day as if I were in a kind of flow state.

Over tea, back at the lighthouse, I learnt more about some of the volunteers. One of them, Kath, lives nearby. Her great-grandfather had worked at Dover Castle, and she recalls that her grandmother had an old hand grenade in her kitchen drawer. Ken is a caver and runs a small museum in Northampton. He'd driven five hours to get here today and would be going home again this evening. Others are discussing caves and mines they've explored. They're all devoutly hypogeous. One man explains how he's tried to like outdoor climbing but that whenever he sees a hole, he just wants to go down it. Another confesses to having enjoyed a recent hillclimbing break in Snowdonia. "What a shame," someone else remarks, as if he's committed a sacrilege.

I was learning about an entire underground culture with its customs and conflicts. After 'operation eyesore' sealed the shelter, local cavers had re-excavated a small entrance. They kept this a close secret, although as entry involved stomach-crawling down the steep top staircase filled almost to the ceiling with rubble, I can't imagine many others would have wanted to join them. When the National Trust announced it was going to clear the tunnels and open them to the public, these cavers, resentful at losing their secret world, had broken in and sabotaged the early clearance work of the volunteers.

Caves and tunnels are two-a-penny around this coast, but sound mirrors are like rare fossils. Unearthing them and exposing their former position on the steep slopes of Fan Bay was the most spectacular part of our work. We excavated around the mirrors to reveal their outline shape, and then the digger moved in, carefully uncovering them with the finesse of a dentist preparing a tooth for a root canal filling. Fully exposed, the mirrors were monumental and elegant, combining these normally opposed qualities in the same way that the term sound mirror combines sight and sound in a synaesthesia all its own. Up close, their concave shape and symmetry is beautiful, not at all sore to the eyes. From a distance, viewed across the sweep of Fan Bay from the north, they look ancient and a bit forlorn, like primitive sculptures an earlier civilisation might have worshipped. Looking out across the Channel from the mirrors, I realised that Fan Bay itself is a vast sound mirror, that the mirrors and the bay echo and reflect each other.

The sound mirrors predated the deep shelter. The first, a prototype of the acoustic warning system, has a fifteen-foot radius and was installed during the First World War. The other, five foot larger, joined it in the late 1920s. Together, they look like a pair of enormous headlights trained on the Channel, watching as well as listening, a perfect figuring of their oxymoronic name. Working up the slope from a group digging at the base of the larger one, their amplified voices drifted up to me like the sounds of the lorries and loudspeakers I'd heard above Dover's Eastern Docks. Standing at the focal point of the mirror (nine feet six inches for the smaller, seventeen feet two inches for the larger) and speaking into it, my words came back amplified. The mirrors are not just quaint, like penny-farthing bicycles, but they work.

Breaking through from inside the deep shelter to the exposed world of the sound mirrors was a special moment, like Cortez as Keats imagined him, 'when with eagle eyes / He star'd at the Pacific'. The moment was recorded by a photograph taken from a drone hovering above us like a kestrel as we gathered around the hole. It felt strangely appropriate that our excavation of this obsolescent form of defence – the first early warning system anywhere in the world – should be captured by the benign operation of a remote killing technology developed by the United States armed forces.

I was wondering again about my pursuit of the remains of war which had now extended to helping in the resurrection and memorialisation of this site. Driving to Fan Bay one Sunday morning for a day's work at the shelter, Elgar's cello concerto came on the car radio. The deep, dark notes of its

opening, Jacqueline du Pré's solo cello beginning its elegiac path through the First World War and its aftermath gave new meaning to the work I would be doing that day. Elgar himself had listened to the sound of artillery from across the Channel while staying at his cottage in Sussex during the war. Standing alongside the sound mirrors that day and looking across the Channel towards the French-Belgian border, I thought of Alf Trevarthen, my grandmother's brother, buried at Bois-Grenier, and particularly of the letter he'd sent home to his brother, Bert, on 10 February 1917:

> *I am writing this from the billet where I have been for 3 days, for a bit of a spell, after coming out of the trenches. I cannot tell you how long I was*

in the line or anything that we were doing, except to say… that things were weird, wild & wonderful & rotten. I have often heard it said that one gets used to shells & bombs but have not seen anyone yet who seems quite at home when they are coming over hot & strong… Well it is still very cold here & the ground is frozen as hard as rock for a depth of about 2 feet. The bread freezes hard now you will find it hard to believe this, but it is a fact… Anything that would shorten the end, so that we could get out of this land of ice and frost would be very welcome… There is not much to write about now so I will finish up hoping you & all at home are well & wishing you "Bon Chance" which is French for good luck & about the extent of my French. Remember me to all.

 I remain Yours truly,

  Alf.

Eleven days later he was dead, killed in action at Armentieres. By the time this letter reached his family in Auckland, they would already have been informed of his death. The first of the sound mirrors was installed in 1917, the year Alf was killed, and from that Sunday morning I came to think of it as his gravestone, his memorial, a marker of the loss and futility of that conflict.

Disinterring the past in the shadows of the deep shelter, I often thought of death, not just Alf's but also the two members of the battery – Harold Aram, a gunner, and Edith Burvil, a NAAFI cook – who were killed here at Fan Bay during heavy bombing in February 1943. In quiet moments during lunch and tea breaks, squatting on the floor, letting the place seep into me, the tunnels had the feel of a mortuary. Or perhaps more precisely, a sarcophagus – a limestone coffin used by the Greeks because of its flesh-consuming properties. Deep in chalk, the air chill, the lighting dim, I felt entombed. Being underground can seem like a prelude to death.

But none of these thoughts stopped me from coming back to work at Fan Bay. I'm a historian by nature, by instinct, and the destruction or suppression of the past appals me. Peeling back a corner of history was justification in itself. And in uncovering the past we were also assisting the future by regenerating the surroundings of the deep shelter and the sound mirrors. Fan Bay is a rare example of chalk grassland, a designated special site of conservation and scientific interest. The spoil and rubble dumped over the entrance to the tunnels and the mirrors had covered the site in rank vegetation alien to its ecology. We were removing this as well as restoring the deep shelter and the mirrors. The surrounds will now slowly regenerate and so the ecological as well as the historical character of the landscape will be restored. This was the double rhythm of the coast that I'd sensed at Plucks Gutter – destruction and loss, regeneration and recovery.

The 'eyesore clearance programme' had been intended to destroy all trace and reminder of both world wars but actually, it created a time capsule that preserved the deep shelter and its signs of human habitation. We found a novel, *The Shadow on the Quarterdeck* (1903) by Major W.P. Drury, pools coupons, a telegram, a needle and thread, a can of shandy. And we found lots of graffiti, mostly names and dates, some dirty stuff where the toilet block had been and a striking example of political graffiti by a soldier who must have been a communist sympathiser: 'Russia bleeds while Britain Blancod'. Blanco was a white substance used to clean soldiers' kit. This graffiti reminded me of the little poem, 'World and Earth', on the wall of the detention block at Richborough, both of them eloquent records of dissent.

Graffiti is also a symptom of underground living. Writing and painting on cave and underground walls goes back so far (at least forty thousand years) that it appears to be a timeless imperative. This urge to inscribe and decorate underground habitations must be an assertion of human presence and selfhood in the sombre settings of depth and darkness. I was to discover this again and again as I went on to explore other underground sites along the Kent coastline. The men of the Fan Bay Battery, the closest defences to France and the most exposed on the front line, had copied their human and indeed Neanderthal predecessors in imitating this practice, recording their lives, their humour and their politics.

The deep shelter had provided relative safety for the battery, a hole in the ground where it made a home of sorts. Our findings, the remains of their time in the tunnels, were the leavings of a small community brought together in haste and duress, abruptly broken up at the end of the war but tight-knit and functioning while it lasted. It's a truism that war often creates a sense of community absent in peacetime, but my experience of working underground with others made me aware that chthonic living encourages this too. Caves in the distant past were places where people gathered together for safety to keep warm, to eat, drink, sing, dance and tell stories. Kathleen Jamie has written in praise of 'the natural, courteous dark', declaring that because of the metaphorical dark, 'the death dark', we've lost the idea that 'dark is good'. The dark can be its own kind of light.

8

DIGGING IN AT DOVER AND RAMSGATE

✝

The first aerial bomb to be dropped on Britain fell in Dover on Christmas Eve 1914. The pilot, Lieutenant Hans von Prondzynski, five thousand feet above Dover Castle, held the joystick with his knees and, lifting the bomb in both hands, heaved it over the side. It missed the castle by several hundred yards and landed in the garden of St James's Rectory in the appropriately named Harold Road, smashing the windows of the rectory and knocking the gardener, James Banks, out of the holly tree he was snipping for Christmas decorations.

It was just five years earlier that Louis Bleriot had made the first cross-Channel flight, bellyflopping onto North Fall Meadow behind Dover Castle. H.G. Wells, in his novel *The War in the Air* (1908), had anticipated that destructive airpower would dominate a coming world war, and Bleriot's flight prompted him to write to the *Daily Mail* warning that Britain's invulnerability as an island nation was now lost. This must certainly have seemed so to the inhabitants of Dover as further and better directed aerial bombing of their town brought fatalities. They began to shelter in Dover's many caves and tunnels, and some with properties backing onto the cliffs started digging their own.

This was an old practice. Dover is honeycombed with caves and tunnels – some natural, others dug – that have been used for centuries for storage, smuggling, refuge and habitation. The 1841 census records several families living in caves on the eastern cliff, and there is a mid-nineteenth-century watercolour of the interior of a cave cottage showing a chimney, windows, a door and a boundary wall of chalk bricks – a strange mix of Victorian domesticity with Robinson Crusoe's enclosure. Chalk, of course, has the great advantage of being easy to excavate. Derek Leach's *Dover's Caves & Tunnels* (2011) describes the nightly procession during the First World War to the caves and tunnels, prams and handcarts loaded with mattresses, pillows, blankets and food. The Oil Mill caves in Limekiln Street with their many high-ceilinged and spacious chambers could hold thousands. Separate caves opening off the main one were designated for men, women and children. Benches lined the walls and electric lighting was installed.

When war threatened again in the late 1930s, Dover's tunnels and caves were reopened, but the Home Office rejected a plan to extend the network, preferring the cheaper alternative of Anderson shelters in people's back gardens. Cost was not the only reason for this. Privatised shelters accorded better with conventional living. Underground spaces were more difficult to supervise, especially in Dover, whose warren of tunnels was incompletely surveyed and mapped. There was concern that, once underground, people would not emerge for work. Subterranean living, free of the constraints of everyday life, was potentially more lawless than life above ground, and indeed during the war, the shelters were used for selling looted property from bombed houses.

By 1941, Dover's underground shelters had been equipped with canteens, heating, WC toilets rather than chemical ones; the floors were concreted and the walls whitewashed. Medical centres had been set up; wardens cleaned and maintained the shelters and were responsible for law and order. But even so, an alternative world to that above ground evolved. Many people ignored the order that caves and tunnels should be vacated during daylight except when there was an air raid. For those whose homes had been destroyed, there was little alternative, but many others preferred the tunnels to their own houses and the Anderson shelters in their back gardens, not just for the greater safety they provided but also for the camaraderie they offered. Recreational activities supplemented the food and medical care that was readily available: cave and tunnel walls were decorated; bingo, communal singing and other entertainments were organised; a traversable underground town evolved. Local people recall making their way to school and work from their shelter without having to emerge into Dover's upper world.

Ramsgate, too, has several miles of tunnels that were used as shelter during both world wars. These were man-made, not natural. During the First World War, more than half a mile of existing tunnel leading to the old railway station at Ramsgate harbour had been converted for shelter. After the war, this became a scenic railway with illuminated wall paintings depicting scenes from around the world. In the late 1930s, with war again in prospect, the borough engineer and surveyor, R.D. Brimmell, conceived an ambitious underground system similar to the network of deep shelters built beneath Barcelona during the Spanish Civil War. At first, the Home Office rejected this plan too, but pressure from the town's mayor, and the alarm created by the German occupation of Czechoslovakia in March 1939, led to its approval. The first set of tunnels were opened later that year by the Duke of Kent.

Unlike Dover's warren of natural underground tunnels, Ramsgate's were planned, mapped and controlled. The scenic railway was closed and its tunnel incorporated into almost four miles of shelter with eleven separate entrances, providing ready access for most of the town's inhabitants. As far as possible, the tunnels followed the line of the town's main roads, using the road surface as reinforced protection against heavy bombing and avoiding legal problems with easement. The world of the tunnels became a simulacrum of the town above, one in which space was measured, calibrated and regulated. Families were given their own separate wooden cubicles; curtains were hung; numbers were added as if the people were living in a street; and signposts were erected.

But for all this careful regulation, underground life was at least exciting for children. Ada Hayton, whose family was assigned to the former scenic railway tunnel, lived opposite the wall painting of a North African scene of sand and camels: 'Life was never dull underground; we would go home to have a bath, wash our hair and get a change of clothes and then back to the shelter.' Rather than going out on the town, they would go out under it. During heavy air raids, she recalled, 'We would hear a dull thud, and know that another bomb had exploded and on those occasions the lights would often go out. This would not deter us little tunnel rats, as we became known. We were more like moles... feeling our way along the chalk walls, we knew every nook and cranny, especially His and Her chemical toilets as we could always smell these because of the Jeyes fluid'. Children ran free.

While I was working with the volunteers at Fan Bay, another group of volunteers were clearing Ramsgate's tunnels. As soon as they were open to the public, I joined a guided tour. We entered at the old railway tunnel from which holidaymakers had once emerged onto the Sands. This vast, cathedral-like space with its high, arching roof had suddenly become home for many local residents after an air raid in August 1940 damaged or destroyed more than 1,200 homes. There's a photo of some of them taken by a *Daily Express* reporter a day or so later, smiling for the camera and holding copies of the *Express*. Churchill visited too. When the air-raid sirens sounded during his visit, he was prevented by the mayor from taking his lit cigar into the shelter. He threw it in the gutter where a local news vendor retrieved it, cut it into short segments and did a brief but brisk trade in selling them. Or so the guide told us.

We followed him for a mile or more along the tunnels. They were uniform in size – six feet wide, seven feet high – and at a consistent depth. Toilet recesses fitted with curtains were placed at regular seventy-five-feet intervals, and there were first-aid posts every thousand feet. Ventilation shafts were positioned at each of the entrances. Life in the tunnels was similarly regulated. Those with permission to sleep down there had an official permit with an ID card recording their cubicle and bunk number in the tunnel and their home address. Dances and concerts were limited to two evenings per week, 7–9.30pm. Alcohol was prohibited and drunks forbidden entrance. Life was much more ordered and patrolled than it could ever have been above ground.

I missed the sense of freedom and spontaneous sociability I'd read about in Dover's less regulated underground world, but I was impressed by the mobilisation, speed and scale of the project. And I was by now feeling at home in chalk, appreciating the beauty of its creamy, scooped tunnels, the elegance of the arches carved into it, how it came to life when flooded with light, its aesthetic. Above all, it was the versatility of chalk I'd come to admire. It was chalk that enabled the tunnels to have been dug so quickly and survive so well. For the Kent miners who did the excavation, it must have been cushy work compared to their usual labour. It had been raining for days when I visited and there were puddles on the floor where water had filtered through the excavated ceiling, but as the guide said, these would soon drain away into the lower depths of the cliffs. A protean substance chalk, one for all seasons.

9

COLD WAR DEFENCES

‡

Down, down, down the lift dropped until, like Alice, I felt as if I were falling right through the earth and might come out in 'The Antipathies'. Deep inside the chalk cliff below Dover Castle, down past the tunnels that had housed a garrison during the Napoleonic Wars, further down past where the headquarters of Operation Dynamo, the evacuation of Dunkirk, had been located, finally coming to rest with a soft thump on the deepest level, home to one of the Regional Seats of Government (RSGs) established in the nuclear world of the early 1960s. I stepped out of the lift into a long-interred office block of large rooms and high ceilings connected by wide, well-lit corridors, spooky because of its very ordinariness. Apart from the lack of windows, I could have been in a city office block thirty floors above ground. This was where the various government departments considered essential for survival after a nuclear attack had been housed: the armed forces, transport, agriculture, health and pension records, the Post Office and the BBC. The plan, similar in a way to Ramsgate's tunnels, was to provide a replica of the official above-ground world to ensure the structure of government survived and could, when it was safe to emerge blinking into a post-nuclear world, be re-established.

There were twelve RSGs scattered across the United Kingdom, bunkers from where politicians, bureaucrats, scientists and communication workers, those deemed too important to be exposed to fallout, would co-ordinate rescue and relief and try to govern a nuked nation. The location of these RSGs was top secret but, of course, news of their whereabouts leaked out. Richard Mabey confesses to having owned a map of one of them during his peace movement days and of hiding it in a matchbox in his clothes cupboard. Even now, Dover's underground site is out of bounds and had only been opened for a couple of weeks to mark the fiftieth anniversary of the Cuban Missile Crisis.

In September 1962, just a few weeks before the Cuban Missile Crisis, a NATO exercise – Fallex – was held over two weekends at a number of Regional Seats of Government. A young naval sailor in Portsmouth, Peter Lindley, was ordered to report to RSG Dover. Arriving at Dover station with no specific directions as to where he should report, and unbriefed as to the highly secret reason for his posting, he started asking people in the street for directions. A policeman, alarmed at hearing the term RSG uttered in public, intervened and took him to the castle. When, a few mornings later, the teleprinter received a message 'nuclear burst in the Channel', his commander had no idea if this was for real or merely an exercise. Worried about the panic he would cause if Channel shipping was alerted, the commander sent Lindley up to check if he could see a mushroom-shaped cloud. Lindley reported that he couldn't.

As the nature of anticipated attack had changed over several centuries, defences had sunk deeper and deeper into the chalk. The White Cliffs remained the nation's defiant outer wall but rather than providing the heights on which proud castles and gun emplacements were sited, they now offered the protected depths inside which the boffins of the nuclear age could plan for survival. Ian Fleming's *Moonraker* (1955), his third Bond novel, catches this shift and gives it a neat twist. It centres on the construction of a nuclear warhead set into the White Cliffs between Kingsdown and St Margaret's Bay, intended, as M tells Bond, to 'give us an independent say in world affairs'. When Bond is shown the eponymous missile by the project's mastermind, Sir Hugo Drax, he is left 'speechless, his eyes dazzled by the terrible beauty of the greatest weapon on earth'. In echoing Yeats's celebrated oxymoron from 'Easter 1916', the passage captures the way in which a drab and reduced post-war Britain might be rescued from the subordinate status to which it has sunk.

Drax is not what he seems but a German leading a team of German scientists (Britain's own are all occupied with rocket testing at Woomera). His covert scheme is to turn the warhead on London. A trial launch with a dummy warhead planned to come down in the North Sea will in fact carry a nuclear warhead and bring the rocket down within a hundred yards of Buckingham Palace. As Bond later realises, 'the Moonraker was a giant hypodermic needle ready to be plunged into the heart of England'. The White Cliffs, the nation's bulwark, has been mined from within and become the base from which its pulse will be stopped. Only Bond and his assistant, Gala Brand, stand between Drax and the end of British civilisation.

At the beginning of the scene in which Bond first realises the danger Drax poses, he and Gala stand on the 'great chalk cliff' near the launching site 'gazing over the whole corner of England where Caesar had first landed two thousand years before':

> *To their left the carpet of green turf, bright with small wild flowers, sloped gradually down to the long pebble beaches of Walmer and Deal, which curved off towards Sandwich and the Bay. Beyond, the cliffs of Margate, showing white through the distant haze that hid the North Foreland, guarded the grey scar of Manston aerodrome above which American Thunderjets wrote their white scribbles in the sky. Then came the Isle of Thanet and, out of sight, the mouth of the Thames. It was low tide and the Goodwins were golden and tender in the sparkling blue of the Straits with only the smattering of masts and spars along their length to tell the true story... As far as the eye could reach the Eastern Approaches of England were dotted with traffic plying towards near or distant horizons, towards a home port, or towards the other side of the world. It was a panorama full of colour and excitement and romance and the two people on the edge of the cliff were silent as they stood for a time and watched it all.*

This set piece, unusual for the Bond novels, is a strongly localised expression of Fleming's love for his country. He knew this coastline particularly well, having purchased 'White Cliffs', the home of his friend Noel Coward on the pebbles under the chalkface of St Margaret's Bay, in 1951. It is curious, therefore, that Bond and Gala can see the cliffs of Margate from their perch above Kingsdown. Tucked around the other side of Thanet, Margate is well out of sight from where they stand. Fleming must mean Ramsgate.

Bond and Gala, who are posing at the rocket site as a security officer and a secretary, are sent by Drax to inspect security at the exhaust pit at the base of the cliff. He tells them that a recent test has melted the chalk, adding: 'Hope we don't burn down the famous white cliffs when we come to the real thing'. This is the first of many references to the fragility of the cliffs and their tendency to crumble. Bond and Gala scramble down the cliff path to 'the wide black mouth of the exhaust tunnel... From the under-lip of the cave melted chalk drooled like lava'. This black hole is a chink in the armour of England's front line defence and the chalk that defends the nation has started to dissolve.

After a skinny dip in the Channel, Bond and Gala dry off at the foot of the cliff. As Bond lies there watching a pair of gulls nesting near the lip of the cliff, 'a great section of the white chalk directly above Bond and Gala seemed to sway outwards, zigzag cracks snaking down its face', and the cliff falls. Only their closeness to the base of the cliff saves them. Looking up, Bond sees 'a jagged rent had appeared in the cliff and a wedge of blue sky had been bitten out of the distant top'. Imagery of mouths and biting run through this scene, most pointedly when Gala, while swimming, looks back 'along the snarling milk-white teeth of England to the distant arm of Dover'. Snarling milk-white teeth is another striking oxymoron, this time Fleming's own, and it captures both the defensive and nurturing qualities of the cliffs but with enough unease to suggest that these teeth might somehow turn and rend the nation they are there to protect – a kind of self-cannibalising.

The cliff fall is the work of Drax, and Bond and Gala are now in deadly contest with the German as they struggle to save the country from nuclear destruction. Held in the Moonraker compound up on the cliffs, they listen to Drax ranting against the hated English – 'Useless, idle, decadent fools, hiding behind your bloody white cliffs' – and boasting that he now has England at his feet: 'Under the very skirts of Britannia. On top of her famous cliffs'.

On the day of the launch, crowds gather along the coast: 'Walmer Beach is black with them... The whole of Kent seems to be out'. Drax's speech, broadcast to the nation just before he presses the button, gloats over the 'arrow of vengeance' he is about to launch, but Bond and Gala have changed the co-ordinates and the rocket comes down not in London but out in the Channel creating a 'great wall of water tearing down' upon the watchers on the shore and the cliffs above.

At the end of Mary Shelley's *The Last Man* (1826), a small group of survivors trapped on the disease-ridden island of Britain gather in Dover hoping to escape across the Channel, only to be confronted by a tidal wave. From the clifftop, they watch the approaching tsunami: 'Would not our little island be deluged by its approach?'. The tidal wave caused by Moonraker prompts a mid-twentieth-century version of a very old, and now a very contemporary, question – will Britain be swamped? Thanks to Bond, the answer is 'no'. The great wall of water destroys the submarine in which Drax and his team are escaping, throwing it out of the water and killing its occupants, but the White Cliffs stand firm.

I'm just inland from Sandwich, walking along the ridge between the villages of Woodnesborough and Ash from where I'd watched the Richborough cooling towers brought down. The path takes me past gardens of late-season sunflowers, luxuriant cabbages and apple trees heavy with ripening fruit. I'm trying to get as close as I can to the bunker.

The nuclear world of the post-war period meant Britain's first line of defence must embed itself deep. A new kind of border was created, an invisible one. A secret chain of radar stations, code-named ROTOR, was dug into the coastline, and it's one of these I'm now looking at, a massive two-storey bunker with thick, concrete walls and a roof of concrete blast slab fifteen feet deep. It's camouflaged by a wide, sloping, grassy mound, which ends abruptly in an imposing concrete retaining wall, like a cliff-face, with steel-plated doors large enough for a double-decker bus to pass through.

The intensified aerial surveillance this underground radar system facilitated brought new risks for the civilian population. One Sunday in January 1954, with the ground temperature ten degrees below freezing, a Meteor Mark VIII jet aircraft, above where I was now standing, lost control and crashed in a field opposite housing on the road between Sandwich and Woodnesborough. The pilot, John Goodwin, just nineteen, was killed and many houses were damaged but, remarkably, there were no civilian casualties. Anne Couchman, a pupil at Sandwich Secondary School, had just gone upstairs to put on her hat and coat for Sunday school when she heard a crash and something heavy exploded through the roof just outside her bedroom. Mr Sydney Burns saw a flying object shooting across the road and into the house of his neighbour, Mrs Pilcher, who had been lighting a fire in her front room when the whine of the plunging aircraft had taken her to the back of the house to see what was happening. As she watched pieces of the aircraft cascading into her back garden, the flying object – the rotor of one of the plane's jets – crashed through the front of her house and embedded itself in the far wall.

In the early 1960s, the ROTOR chain was superseded by new kinds of radar and the Ash bunker was sold to the Civil Aviation Authority for £1. But the 1981 Defence White Paper, preoccupied with the threat of Soviet Russia, emphasised the need to protect air defence radar and communications systems from the targeting power of modern long-range weapons. The bunker was taken back by the RAF and re-engineered to withstand a hit from a twenty-two-kiloton thermonuclear bomb. Its reconstruction took several years and brought many complaints from the surface-dwelling inhabitants of Ash because of damage to their homes caused by the excavations.

But for all the urgency of the Defence White Paper, RAF Ash was never fully operational. Diesel generators and reinforced concrete roof panels remained on the site unpacked and unused in defiance of the stated aim of the Defence White Paper 'to get the best value from the resources we spend on defence'. Cobbett's complaint about 'wasted expenditure' was still relevant as the detritus of defence continued to litter the Channel coast. After the collapse of the Soviet Union brought an end to the perceived danger from nuclear attack, RAF Ash was slowly run down and then sold to its present owners, a computer security and data storage firm.

I wanted to see inside the bunker. By now I'd developed a taste for the underground and understood the obsession with exploring holes in the ground that the volunteers at Fan Bay had introduced me to. I wrote several times to the company asking if I could visit the bunker but never got a reply. I knew it well from the road, a compound surrounded by formidably high wire fences with an overhang of coiled barbed wire, a double-gated entrance and festooned with warning signs. It looks like one of those prisoner of war camps in Second World War films. Whenever I cycled past and slowed to peer in, a menacing black dog would come to the wire and growl.

So I'd taken the path along the ridge to get a view of the compound and its installations from above. As I scanned the place and scribbled some notes, a man appeared on top of the grassy mound that sheltered the bunker and stood there staring hard at me. There were just the two of us, me standing exposed in the middle of a ploughed field, he on top of the bunker, eyeballing each other, a belated cold war encounter, absurd and uncomfortable. I lost the staring contest and, disconcerted, out-spied, feeling rather foolish, I moved back up the hill and rejoined the footpath.

10

CURIOS AT EASTRY AND MARGATE

‡

Andrew Meachan, the owner of 'Beckets', pulled back the heavy wooden trapdoor in his garden and led me down a steep flight of Victorian brick steps into the darkness. I was becoming familiar with the feel of going underground, the dead-still, slightly clammy air mixing with the dry smell of chalk and the strange sensation Kathleen Jamie has described of entering a body and moving through its chambers. The tunnels were spacious, with high, arched ceilings and straight-cut walls creating long corridors narrowing into the distance, picked out by our torches. Shorter tributary tunnels branched off the main routes, eerie galleries momentarily lit up by our raking torch beams. Once more, I thought of Alice in her underground wonderland (Lewis Carroll's original title was *Alice's Adventures Under Ground*). But unlike the curious Alice, left to pick her way as best she can through the strange world of Carroll's imagination into which she falls, I had a guide. Andrew knew the tunnels as well as he knew his own house.

I was in the village of Eastry, just inland from Sandwich and a couple of miles south of Ash. It dates from Roman times and in the seventh century became the Saxon capital of Kent. The royal palace of the Saxon kings is thought to have been located on the present-day site of Eastry Court, next to the parish church. It was here, around 670, that Egbert, King of Kent, had his young cousins, Ethelred and Ethelbert, murdered and buried in the King's Hall.

Eastry has a labyrinth of underground tunnels, the entrance to which is from the garden of Andrew's seventeenth-century home. They have been used for refuge – Thomas Becket is said to have sheltered here while escaping to France in 1164 – secret worship in times of religious persecution and for smuggling. In the nineteenth century, a local bricklayer and lime burner began extracting chalk from the tunnels. Quicklime produced from the burning of chalk was widely used in Kentish agriculture and brickmaking. It gave a distinctive yellow colour to the brick houses in this area, including my own. Chalk burning ended during the First World War, after which the only commercial use of the tunnels was a short-lived enterprise in the 1960s which opened them to the public. Eastry Rural Council closed the venture after someone was lost in this warren for ten hours.

Stumbling along in Andrew's footsteps, it was easy to understand how someone might have wandered lost for hours. If Andrew had slipped away into the dark, I'd have had no idea how to get out. One of the side tunnels came to a blackened end where lime-burning had taken place. Another side tunnel was blocked off by a rendered wall with an elaborate fresco of the murder of the Saxon princes of Eastry. There were sculptured heads – a Roman, a pagan god – and a large painted figure of a Viking warrior with a legend recording it had been done in 1911. The longest and straightest of the tunnels was peppered with bullet holes from when it had been used by the Home Guard for rifle practice during the Second World War. It would have made a perfect setting for an episode of *Dad's Army*. Andrew told me that scenes for a Nazi zombie horror film had recently been shot down here.

There were graffiti everywhere, hundreds and hundreds of scrawled names – the earliest I could find dated 1851, the most recent 2013. Traditionally, these tunnels were a place of festivity, an underground playground for the village. On Fair Day, 2 October, the workings would be illuminated with candles and the villagers given free run. The older graffiti in the tunnel dated from this time, a record of the annual infringement of this normally sealed underworld. Here were more signs of how underground chambers and the alternative world they provide encourage carnival, a topsy-turvy otherness inimical to regulation and control, freeing the individual and the community for a day from the prohibitions and constraints of everyday existence.

Just as caves and underground chambers seem to compel inscription, so too do they generate stories and legends, as if their dark, uncanny emptiness must be filled with human narratives. Mrs Moore experiences this need when she visits the Marabar Caves in E.M. Forster's *A Passage to India* (1924) and is overcome by 'a terrifying echo' that murmurs in her ear, 'everything exists, nothing has value'. This is a noise without meaning, a confrontation with the void that wipes out her belief in 'poor little talkative Christianity' and the narratives of family and empire it sustains. Others fill the caves with their own desperate stories – Adela's accusation that Aziz has assaulted her, the white man's narrative of the innate depravity of the Oriental – but Mrs Moore loses her belief in the possibility of belief. Caverns like the Marabar are 'measureless to man', filled with an emptiness that can only be withstood by stories that attempt to protect us against the loss of identity and meaning they threaten. Alice does this too, trying to interest the inhabitants of the underground world of her dream narrative with stories of her above-ground life, only to have them interrupt, misunderstand, walk off or fade away.

I'd become fascinated with the idea of underground worlds as an archive of secret histories, the unconscious of the East Kent coast, sites of things forgotten, repressed or invented. Like shipwrecks on the Goodwin Sands, such places have repeatedly been found, lost and recovered, opened and closed down, generating and multiplying the stories they harbour.

Margate has several vivid examples of this. The documented history of 'The Cave of Vortigern' dates from 1798 when it was uncovered by a gardener working in the grounds of Northumberland House. According to one account, he fell into the cave and died from his injuries. The owner of the house extended the diggings and commissioned a local artist to decorate the chalk walls with carvings and paintings. In the 1860s it was named and themed 'Vortigern's Cavern' and opened to the public as a commercial venture. Vortigern was the British king said to have granted Thanet to the Saxon invaders, Hengist and Horsa, and the proprietors marketed the underground chambers – a pair of ovoid pits and a deep shaft connected by a network of galleries – as dating from this time. In fact, modern surveys indicate that it was originally just another chalk mine excavated around the turn of the eighteenth century.

One of the myths about the cavern was the existence of a 'Lido tunnel' connecting the diggings at Northumberland House, which is several blocks inland from the seafront, with Margate Sands. Boats laden with contraband were said to have been floated at high tide along this tunnel to the caves where the goods were stored. Hypogeous storytelling often involves separate underground cavities being linked together to create networks – a continuous underground world rather than an isolated subterranean space. There is a chalk cave at the south end of St Margaret's Bay, for example, which was said to run underground all the way to Canterbury, eighteen miles away. Excavation has found no trace of a 'Lido tunnel' in Margate.

'Vortigern's Cavern' was an attempt to profit from the burgeoning seaside holiday market. It failed, as have many other attempts to exploit underground worlds along this coast. The Margate Shell Grotto, though, is an exception. When I last visited, on a cold, damp Sunday morning in February, a short queue was already waiting in the street for the grotto to open. It lies beneath a house in a modest, inconspicuous, suburban hillside of late Victorian terraces. Dropping down a short flight of steps with rough chalk walls, you enter the grotto, a serpentine passage interrupted by a domed rotunda and ending in an altar chamber – a peristaltic journey of just over a hundred feet past walls lined with an estimated 4.6 million shells. Unlike the other caves I'd explored or researched, whose void required filling with images and stories, the Shell Grotto was replete on discovery with a mosaic of intricate visual patterns and images.

The shells have faded to a monochrome grey, like weathered concrete. The symbols and patterns they form are difficult to discern at first but slowly resolve into an intricate and elaborate iconography: representations of plants, flowers, trees, turtles, snakes, gods, sexual organs, birth, stars, suns and moons. Unlike the classical themes and decorations of the grottos that became fashionable in the eighteenth century and were decorated with exotic shells, these shells are a medley of local winkles, cockles, mussels, oysters and whelks, most of which are thought to have come from Pegwell Bay.

The complexity of the designs and the delicate skill of the whole conchological mosaic prompt questions: who made them? when? and why? There are any number of theories about the grotto, encouraged by inconclusive results from carbon dating. The case has been made for it being a Mithraic temple. The cult of Mithras was popular among Roman soldiers, but its defining image – the god, Mithras, slaughtering a bull – is missing. The iconography of the grotto is entirely to do with fertility and the natural world. Others have connected the grotto with the Knights Templar (the foundations of a Knights Templar church have been excavated on Dover's Western Heights). Phoenician, Minoan, Hindu, or Tudor origins have also been suggested. A Margate historian, Mick Twyman, noting how at the summer solstice the sun shines into the grotto through its dome and fills the whole tunnel with light, calculates that it was constructed around 1141. The precision of his dating is based on the creep of one per cent in the angle of the equinox that occurs every seventy-two years.

The grotto has its ghost, as all such spaces should – 'The Lady of the

Temple', aka 'The Blue Lady' – who walks the Serpentine Passage. The grotto has also been used for supernatural occasions. There's a photograph in one of the display cabinets of a seance in the Altar Room in 1939. A white-faced medium stands among a circle of seated, fur-coated women, their faces mainly in shadow, a tonal contrast that intensifies the spectral quality of the image. I wondered if the imminent prospect of war had revived a practice that had become so common during the First World War.

The eclectic iconography of the Shell Grotto – 'Crazy / curious muddle of Christ and kabbalah / the rosy cross, lotus and phallus, a sign / for everyone', as Sally Minogue describes it in her poem 'Shell Grotto' – seems to exceed all possibility of interpretation. Even the story of its discovery is contentious. The official account is that it was unearthed in 1837 by gardeners working in the property of a Mr Newlove, who then opened it to paying visitors. One of his children, however, in old age, claimed it had been discovered two years earlier by she and her siblings while playing in the garden of the house which their father was then only renting. It's possible, therefore, that the children's initial discovery was kept secret by the family until their father had secured purchase of the house and garden and that the subsequent discovery by the gardeners was a staged or fabricated event.

Sonia Overall's novel, *The Realm of Shells* (2006), makes delicate play with this disputed history and dramatises the range of different meanings the underground domain of the grotto has for the children who discover it and the adults who wish to exploit it. At the end of the novel, Newlove, an evangelical Christian ruined by his compromised efforts to commercially exploit the grotto, has retreated underground into 'the pit of his own making', his private hell, where he rails against the God who has let his enemies prevail: 'you are not here – you are not everywhere! This is a place of devils'. His despair recalls the terrifying emptiness of the Marabar Caves and echoes the title of David Seabrook's Margate-centred book, *All the Devils Are Here*. Underground caves can do this to you.

11

THE EAST KENT COALFIELD

‡

Beside a roundabout on the A258 between Sandwich and Deal is a prominent sculptured figure, 'The Waiting Miner'. Crouched on one knee, naked from the waist up, the miner faces inland across fields of vegetables towards low hills; the coast is less than a mile away at its back. At first it seems incongruous. No one now thinks of Kent and coal. There is nothing in the surrounding landscape of low-lying arable farming and marshland to suggest coal mining, but in fact 'The Waiting Miner' is looking directly at the former site of the Betteshanger pithead just a few hundred yards the other side of the roundabout. And the gently sloping, lightly wooded hill to its back is the reclaimed tip site of the Betteshanger pit, now a country park. The sculpture is the only obvious trace of a vanished industry that dominated the area in and around Deal for sixty years until the colliery was shut down in 1989.

Around the base of the bulky plinth of 'The Waiting Miner' are columns of names of the 'Men Who Lost Their Lives During the Life of the Kent Coalfield 1896–1989'. Deaths at Shakespeare Colliery (1897–1912), Tilmanstone Colliery (1909–1981), Chislet Colliery (1915–1964), Betteshanger Colliery (1927–1989) and Snowdown Colliery (1907–1987) totalled 165. In fact, research by former Snowdown miners has uncovered further deaths and the names of these men are now recorded on a pair of granite stones in a recently established memorial garden in their pit village, Aylesham.

The oldest miner to lose his life was sixty-nine, the youngest, two lads of fourteen: D. Henderson, Snowdown, 1941 and A.W. Moore, Tilmanstone, 1936. I came across the grave of Arthur William Moore one day in the overgrown churchyard of All Saints, Waldershare, near Tilmanstone, one of those abandoned churches that are scattered around this part of Kent, as redundant as the coal mines. His gravestone reads: 'Accidentally killed at Tilmanstone Colliery 31st Dec 1936. Aged 14 years & 7 months'. The grave is shared with his parents, Winifrid Maud Moore, who died in 1977 aged seventy-five, and William Moore, a Tilmanstone miner himself, who died four years later at the age of eighty-nine.

The church is small and its interior plain, except for the chapels to either side of the chancel. One of these has a large, solid tomb chest erected by Sir Peregrine Bertie to his wife, Susan, after her death in 1697. Their effigies lie together on the chest, hands clasped. I thought immediately of Larkin's poem 'An Arundel Tomb', but unlike the covert handholding of his earl and countess, Sir Peregrine and Susan's is overt, public and very touching. Larkin's aristocratic couple has been transfigured by time to a 'stone fidelity / They hardly meant'. The fidelity of the couple in All Saints, their 'final blazon', is deliberate, tender and quite unexpected in the chill of the unused church. The other chapel is filled to overflowing with an enormous three-tiered monument, like a fantastical wedding cake, to Sir Henry Furnese (d.1712). This pompous erection made of four kinds of marble and flanked by statues of women mourners and cherubs, an assertion of the wealth and power of the local landowners, expresses perfectly the neo-feudal social relations of bygone rural Kent. The grave of Arthur William Moore, on the other hand, is a small monument to the depredations of Kent's short-lived industrial phase. But there is also something redemptive in Sir Peregrine's elaborate memorial to his love for Susan, a reminder, to quote the final line of Larkin's poem, that 'What will survive of us is love'. Winifrid and William Moore must have lived with this knowledge for over forty years.

The list of dead around the base of the sculpture of 'The Waiting Miner' makes it seem like a war memorial, and indeed two of the fatalities at Betteshanger were the result of a bombing raid on the mine in 1942 that left many others trapped underground. Coal mining was closely associated with war, the skills of the miner being employed in digging trenches as well as shelters in the chalk and bringing up the coal to fuel the industries that sustained the war effort. This was captured by Wilfrid Owen in his poem, 'Miners'. The 'I' of the poem hears a sigh from the coal in their hearth:

I thought of some who worked dark pits
Of war, and died

Digging the rock where Death reputes
Peace lies indeed.

It is not one of Owen's best poems, but it is distinctive in ranging beyond the usual subject matter of First World War poetry to draw a connection, both figurative and material, between the industries of war and mining and those who suffer from the exploitations of both, the soldier and the miner. Henry Moore knew this. Around the same time as his drawings of Londoners sheltering in tube stations, he produced another underground series, of miners labouring at the Wheldale Colliery where his father worked.

The discovery of coal along this strip of coast was an accidental result of the earliest attempt to build a Channel tunnel. Work on this had begun at Shakespeare Cliff at the south end of Dover in 1880, but the government soon called a halt, concerned that a direct connection to Europe would open a new border to defend. A new company, the Kent Coalfields Syndicate, purchased the workings and sunk a shaft. The first loss of life in the Kent coalfield occurred in 1897 when a team of fourteen men working at Shakespeare Colliery was engulfed by water shooting up into the shaft from below. Eight of them drowned.

Over the next thirty years, more than forty boreholes were sunk in the area. Government plans in the early 1920s predicted eighteen pits in East Kent and the need for fifty-five thousand houses for miners to be provided in a number of projected new towns. Iron ore had also been discovered in the area and an export trade with associated steel industries was planned for the First World War port of Richborough. If all this had come about, East Kent would have resembled South Wales. In the event, only the five collieries named at the base of 'The Waiting Miner' ever became productive. Several speculative ventures went bust and the grand plan failed to materialise.

Arthur Conan Doyle was one of those who lost heavily in what he called a 'wildly financed and extravagantly handled' enterprise. Before investing in the coalfield, he'd descended a thousand feet through the chalk to inspect the coal in situ. Coal it was, but as he wrote to his sister Mary, it proved to be incombustible. He recounted that 'when a dinner was held by the shareholders, to be cooked by local coal, it was necessary to send out and buy something that would burn'.

The economy of the East Kent coast at this time was mainly pre-industrial – fishing, farming and the holiday trade being its mainstays. There was very little ready or experienced labour in the area and the bulk of those working the new pits had migrated from coalfields across the country. By the end of the 1920s, just a couple of pit villages – Elvington to house Tilmanstone miners and Aylesham for those working at Snowdown – had been established. Purpose-built to accommodate the incomers and their families, they were geographically isolated and socially distinct from the farming villages and hamlets that lay around them. Betteshanger miners were concentrated in the newly built Mill Hill Estate on the edge of Deal.

Mining families brought with them different accents and dialects, social customs and work practices – unionisation in particular. The traditional insularity of mining culture was reinforced by the way the incomers were regarded as aliens and by the settlement patterns. Miners from the same region would often gravitate towards the same pits in the new coalfield. Betteshanger, for example, was predominantly Welsh but with miners from Scotland and the north of England as well. Antagonism between these different national groupings was common at first but over time each mining settlement developed a distinct and cohesive identity. The short history of Kent mining inscribed at the site of 'The Waiting Miner' records that its communities were often referred to as 'tribes'. These strong local identities could also create tensions between the four pits that made up the Kent coalfield.

Betteshanger opened later than the other mines and most of its workforce arrived in the wake of the 1926 General Strike – out-of-work miners, many of whom had been blacklisted in their former pits for militancy. One such man, Gerry Quick, walked from Merthyr Tydfil, sweeping streets in Newbury and taking fieldwork as he journeyed, arriving in the pit yard a month later, malnourished and with bleeding feet. Similar journeys were made from other parts of Britain, mainly on foot or bicycle. Jack Dunn, sometime Kent area Secretary of the National Union of Mineworkers (NUM), described the influx of men: 'We had Scots, Durham, Lancs, Staffs, Derbyshire miners. It was like a cocktail, they all had a different dialect, and all gave a different name to the same tool. They had different habits and cultures, so it was difficult to get a quick coalescence'. Their experiences, many of them recorded by the jazz singer Gina Harkell in the mid-1970s, have much in common with those of refugees crossing national borders in search of work and security today, and so too does their reception.

The reaction of many inhabitants of Deal to the sudden arrival of 1,500 miners in their homogeneous seaside town was similar to the alarm provoked by present-day stories of 'floods' of immigrants threatening the nation's identity and resources. Signs in lodging houses, pubs and cafes declared 'No Miners', and shops advertised 'miner's bacon' – leftover scraps of meat and fish. At first there were no pithead baths at Betteshanger, and the miners would return to Deal at the end of their shift black with grime. One of them who spoke to Gina Harkell recalled: 'They seemed to regard miners as some kind of weirdoes – a man who can go and grovel in the earth, he must be lacking. He doesn't deserve any sympathy and they despised us'. A headline from the local newspaper at the time catches this: 'Man and Miner Fight in the Street'. By 1930, the Mill Hill Estate of 950 houses had been established; sports facilities and a social club followed; and a distinct identity, similar to that of Aylesham and Elvington, but without the self-containment that was the hallmark of those more isolated communities, had formed.

The feeling of being strangers in an unwelcoming social environment must have contributed to the militancy of the Betteshanger pit. The main reason though was its concentration of union activists from the 1926 strike. Some blacklisted miners arriving at Betteshanger took on assumed names to avoid being excluded from the new pit. They brought with them a political culture described by Jack Dunn as being in 'the best Welsh tradition': 'we had a library in the miners' club and… people used it regularly every day, they'd come in and read Hansard, follow the debates on the industry'. This resulted in a politicised and militant workforce with a strong communal sense of self-sufficiency that was carried back to the surface each day and which defined the above-ground lives of miners and their families as well.

This made Betteshanger a difficult pit to control. An early example of its intransigence was a two-month strike in 1938 over pay, working conditions and the treatment of pit boys. Arthur Moore's death had occurred just a year earlier. About 1,600 miners marched from Mill Hill to the Deal Labour Exchange and handed themselves into the local workhouse, a tactic designed to embarrass the employers and local authorities. One pit boy remembered being sent to Minster where he and the others extended their strike action to include the tasks they were set at the workhouse. He recalled his time there with great pleasure: 'I think it was June, it was lovely. So we walked home and got our bikes and went back there. We drove the warden mad; we used to go riding round all the cherry orchards and fruit orchards. We even went down to Birchington singing with a cap, to get money to go out in the evening'. It's a shame that Arthur Moore missed the fun.

When the Second World War began, the Kent pits, the only source of coal south of the Thames, were in the front line. Betteshanger suffered several direct hits and was always in danger from stray bombs offloaded by German planes making their way back across the Channel after raids over London. An elderly man I got talking with when visiting a local display of mining photos remembered regular bombing raids on the mine – 'Every Sunday morning,' he told me. Many of Deal's inhabitants were evacuated and this intensified the isolation of the mining community. The wife of a miner who arrived in Deal in 1941 described it as 'a ghost town'.

Early in 1942, the pit went on strike over the allowances being paid to work a particularly wet and difficult seam. This breached wartime emergency regulations and was disowned by their National Union which echoed the patriotic rhetoric of the mine owners and the government: 'Strikes or lockouts cannot be permitted when the enemy are at the gate'. The government and the owners decided to prosecute: three union officials were gaoled, and one thousand men were given the option of paying a fine or being sent to gaol with hard labour. All but nine of the miners refused the fine and the authorities were faced with having to find prison places for the rest. Support for the strike among miners spread as far as Scotland and the government, worried that other pits would follow suit, backed down. The union officials were released and the fines imposed on the other striking miners waived.

Betteshanger's reputation of striking against the national interest in time of war was never forgotten and was used against the pit as closure threatened in the 1970s and '80s. It was also a factor in the neglect of the mining communities in the years immediately following its closure. Betteshanger – the closest mine of all to the Channel coast – was the dark underbelly of the White Cliffs, undermining that symbol of defiant nationhood, challenging the idea of one nation at its very border.

I've never been down a mine and the opportunity to do so has now passed. But George Orwell did, the same year in which Arthur Moore was killed. His classic account of that experience in *The Road to Wigan Pier* (1937) describes the world of the miner as being like his mental picture of hell: the noise is deafening; you cannot see far because 'the fog of coal dust throws back the beam of your lamp'; the heat is suffocating; and the coal dust which 'stuffs up your throat and nostrils and collects along your eyelids' is choking. The work is an assault on every part of the body. The daily journey from the pit bottom to the coal face is on average perhaps a mile and can be as many as three, bent double much of the way through thick dust, with jagged shale and often water underfoot, 'mucky as a farmyard'.

Orwell's account would be generally true of Kent mines as well, although every mine had its own character and even within each one conditions varied according to the seam being worked. Snowdown was very deep, and poor ventilation made it probably the hottest pit in the country. Miners there would work in women's knickers because they were lighter and cooler. Snowdown and Betteshanger were consistently wetter than Orwell describes because of vast underground lakes at that depth. And his account focuses on the work of the fillers at the coal face. Arthur Moore was killed while working in haulage, conveying coal from the workings to the foot of the shaft.

The keynote of Orwell's description is of the mine as a world apart, one which he says most people would prefer not to hear about even though the above-ground world depended so heavily on its labour. 'Practically everything we do', he wrote, 'from eating an ice to crossing the Atlantic, and from baking a loaf to writing a novel, involves the use of coal, directly or indirectly'. Hence the common description then of coal as 'black gold'. But this separation of upper and nether worlds never applied to the miners themselves. Pit and village were as one.

Aylesham, with its attendant colliery, Snowdown, was the most perfectly integrated example of this symbiotic relation. This completely new, purpose-built town was designed by Patrick Abercrombie, later famous for his replanning of London after the Second World War and as instigator of the New Towns Movement. Although he sited Aylesham a mile or so distant from its colliery, unlike traditional mining communities where village and pit often ran into each other, he modelled its layout on the headgear of a mine, its centre of public buildings and shops being in the shape of a pulley wheel with two lines of houses running away from it as if supports of the winding gear. The working world of the pit was imprinted on the very shape of a town spatially defined in terms of its economic function.

It's remarkable how a planned town with a precise year of inauguration – 1927 – rapidly became a cohesive and self-determining community and has, to a considerable degree, survived as such. The miners and their families who founded Aylesham brought their wider culture with them. A brass band, a male voice choir, pigeon clubs, whippet racing, colliery sports days and sports teams were soon established. A distinctive Aylesham accent even developed. Like many migrant communities, the culture they brought with them intensified rather than diluted in their new environment. And the continuing strength of this culture has helped the community of Aylesham survive the closure of its mine. Residents three generations away from any direct connection with the mine still describe themselves as being 'from a mining family'. The town's heritage building has an extensive paper and electronic archive of its past where former miners gather on a Wednesday morning to work on the archive, plan activities and chat over a cup of tea. It is hard to think of another workplace extinct for so long that continues to sustain a whole community in this way. Usually when the work goes, a town dies.

Betteshanger was never as self-contained as the Aylesham community. Migrants in a town that did not welcome them, the miners on the Mill Hill Estate formed more of a ghetto than a community on the Abercrombie model, isolated but surrounded. Industrial relations at the pit and social relations in the town mirrored each other and fed the militancy that resulted in sustained actions of a kind not found elsewhere in the East Kent coalfield. Stay-down strikes for example. In 1960, when 120 young miners were made redundant, almost four hundred older miners occupied the pit and stayed down for eighteen days. Mill Hill rolled into action. Blankets, clothing and food were sent down to the men; a daily postal service was set up; a skiffle group formed below ground to keep everyone entertained; and the Southern Counties Light Heavyweight Boxing champion had his skipping rope sent down to help him keep fit. Some of those above ground went back to their coalfields of origin across the country to raise support. At the end of the stay-down, the men emerged to the greeting of the Betteshanger brass band. Some, not all, of the jobs were saved. This stay-down tactic was repeated in 1984 during the prolonged national coal strike when, as Betteshanger lay idle, management announced that it was full of gas and would have to close. In response, striking miners occupied the pit and proved the gas story a lie. The National Coal Board was forced to withdraw its claim and concede that the pit was gas-free and in good condition.

Betteshanger was the last pit in the country to return to work after this bitter year-long national strike, and its action throughout this period was sustained by the Mill Hill Estate. Miners' wives and partners formed support groups which raised funds, organised food collections, arranged holidays for the children and travelled to Belgium and Germany to win support for the strike. The pit and its community – two worlds as one – managed to sustain itself for more than a year while challenging a capitalist system in the process of de-industrialising and to inspire counter-narratives, alternative visions of human society, as dramatised by the 7:84 theatre company's play *Garden of England*, based on the experience of Liz French, whose husband Terry was imprisoned for five years for picket line offences.

The 1984 strike, however, was the beginning of the end for Betteshanger, the other Kent pits and the whole mining industry. Government documents released in 2014 confirmed that the Kent coalfield had been earmarked for closure before the strike. Co-ordination between the government, the National Coal Board, the security agencies (MI5 tapped the phones of NUM leaders) and the police frustrated tactics such as the use of flying pickets that had proved successful in previous strikes. The crossing of county borders was subjected to similar control as that of crossing national boundaries. Pickets from Kent were turned back at the Dartford Tunnel and threatened with arrest. A convoy of food from Holland for the miners and their families was impounded at Dover.

Early one tranquil summer morning during this period, I drove along quiet, winding lanes towards Tilmanstone to join a picket line. With me was a friend, a fellow member of the Association of University Teachers which was supporting the miners. Suddenly, from out of nowhere it seemed, our way was blocked by a small posse of policemen. They questioned us, searched the car and told us to turn back. Richard, an academic lawyer who had recently written a piece for the *Guardian* on the illegality of the blockade at the Dartford Tunnel, was well primed to argue, but the police warned that if we challenged this obstruction to our freedom of movement, we'd be arrested. Nothing could have made clearer Margaret Thatcher's description of the miners as 'the enemy within'.

Betteshanger, the last surviving pit in the East Kent coalfield, closed in 1989. The last death at the mine occurred just a few months earlier when Geoff Almond was buried under tonnes of slurry while cleaning a conveyor belt. Some men found employment digging the Channel Tunnel, but union leaders and other activists were refused work there, the blacklisting practices of the 1920s now extending beyond the mining industry. Others found employment doing track maintenance on London Underground. Faltering efforts at regeneration came and went. The now defunct South East England Development Agency (SEEDA) tried unsuccessfully to set up a business park on the former colliery site. This is the empty space that 'The Waiting Miner' faces. Where the pithead, the winding gear and the colliery buildings once stood, the ground has been cleared and levelled. The only evidence of a coal mine having been there is two small slabs marking the position and depth of its former shafts: 'No 1, 687.65m'; 'No 2, 739m'. One building remains, 'Almond House', but there is nothing to explain the significance of its name.

At the back of 'The Waiting Miner', however, Betteshanger's spoil tip has been reclaimed and regenerated to create a country park. The very stuff of the reserve, shale from the colliery that has been mixed with recycled green waste and fertiliser, is a reminder of the history you are walking on. Pine, birch and alder, trees that tolerate poor, well-drained soil, were planted to stabilise the shale and create woodland. Scrubland was planted to encourage field birds. Meadow pipit, linnet, goldfinch and ground-nesting birds such as plover and partridge inhabit the park, and kestrels and sparrowhawks hover overhead. There are paths for walkers, a tarmac cycle track and trail routes for bikers, an adventure playground for children. The lookout point on the eastern slope offers a sweeping view of the coastline and its immediate hinterland. In winter, when the wind comes down from the north-east, this exposed ridge becomes the most eye-watering, jaw-numbing, bladder-gripping point along the coast, the physical discomfort somehow appropriate for a site where local miners scavenged for fuel when on strike and beneath which many lives were lost.

Kent's mining past has continued to fight back. The siting of 'The Waiting Miner' at the entrance to the park was the result of a long campaign by former miners, and the name of the park, Fowlmead (a bowdlerisation of the original name of the marshland, Foul Mead), was changed to Betteshanger in recognition of its mining past. And in 2022, the Kent Mining Museum opened and has become the centrepiece of the park. The experiences and memories of former miners, their wives and families fill the museum. The extensive archives gathered at mining heritage centres in Aylesham and Elvington provide the documentary foundation of the stories the museum tells. Plans for an accompanying Green Energy Centre producing renewables in the form of green biomass energy on the site have, however, been abandoned because of the collapse of the original financial backers of the development of the park, Hadlow College – a story that recalls Conan Doyle's description of the early mining enterprise in which he lost so much money. Unfortunately, the opportunity to revive in sustainable form the tradition of energy production of the Kent coastline has, for the moment, been lost.

A recurring theme of my explorations has been the never-ending process of loss and recovery, destruction and regeneration, whether natural, as with the geography of the coastline itself; human, as at Fan Bay; or indeed personal, as I tried to establish this littoral as a place where I felt at home. Exploring the defences of the Kent coast as the history of the First World War was being recounted and memorialised had made me reflect on how things past are remembered, framed and narrated. The story of the Kent coalfield – how it had been closed, almost erased and was now being rewritten – had raised this again.

The history of 'The Waiting Miner' focused this question. The statue was originally sited at Richborough Power Station where much of the coal from Betteshanger went. When Richborough was decommissioned, it was moved to the seafront in Dover and positioned next to the former National Coal Board (NCB) offices where it sat rather oddly looking out across the Channel. Or so it seemed to the 'Move the Miner Committee' which campaigned to have the statue shifted back to the coalfields. As one former miner put it: 'It's facing the Channel so it's got its back to us, so there is disrespect really'. And there was no love in the East Kent coalfield for the NCB.

But the question of where to reposition the statue brought back old rivalries. Former Snowdown miners wanted it in Aylesham where it would have been owned by the town, perpetuating the inwards-looking character of mining in East Kent. Those from Betteshanger preferred the roundabout on the A258 that cut between the former colliery and its spoil tip where it could be noticed by passing motorists and walkers – a more outwards-looking affirmation of the mining tradition.

Aylesham, in fact, already had three sculptures of its own marking the village's mining past. The first of these was a pulley wheel salvaged from the colliery when the winding gear was demolished in 1987 and laid, just as it was, on a stone base in the market square by the town's oldest living miner, ninety-seven-year-old Jack Christian. The second, 'The Aylesham Phoenix', a piece of conceptual sculpture representing the upright body of a phoenix encased in sheets of flame, was externally commissioned and placed outside the former secondary school whose closure, like that of the mine, had been deeply resented. It was so disliked by the community that they designed and commissioned an alternative sculpture of their own, 'Payday at Snowdown Colliery', sited less than a hundred yards from 'The Phoenix', outside what is now the 'Miners Way Business Park'.

This is a larger-than-life arrangement of a miner and two children in front of three trucks of coal. Its realism is a standing rebuke to the abstract Phoenix, endorsed by a detailed information board explaining what the sculpture illustrates:

> *Children rarely went to Snowdown colliery but as a treat on school hoildays* (sic) *they sometimes went with Dad to the pit canteen for the traditional miner's dinner of pit pasty, peas and mashed potatoes and if he was on the afternoon shift they would bring dads* (sic) *weekly pay home to mother.*

The sculpture is both scrupulously naturalistic and sentimental. The miner is wearing his pit helmet, his right hand supportively behind the back of his daughter who carries his lamp; his snap-can for his bread-and-dripping is in his left; his young son is on his shoulders. The man's face is set, his eyes fixed and determined. Clothes and other details set the figures firmly in the 1950s. The whole configuration is an idealised picture of a mid-twentieth-century working-class family (the absent mother at home), underlined by the narrative of a former way of life that the ensemble is celebrating. It is there to set the record straight, to fix and preserve it, to clarify the obscurity of the phoenix.

The authenticity of the sculpture is unquestionable. This was the village's response to what they saw as yet another betrayal, the phoenix being the final insult to the injuries of a shutdown mine, a closed school and a community left struggling to survive. In the face of the neglect suffered by Aylesham after the closure of its pit, the image of rejuvenation that the phoenix was intended to convey was bitterly inappropriate. 'Payday at Snowdown Colliery' responded by insisting on the value and integrity of what had been lost, an assertion of the dignity of the miner and his loyalty to those who depended on him. This, however, has left it frozen in the moment of its commissioning when resentment at closure, neglect and the imposition of an unwanted and disliked art object on the village was at its strongest. Several decades later, the work has become a passive monument, its legibility and eloquence confined to the moment and circumstances of its installation.

'The Waiting Miner', by contrast, slips the net of its originating history. In this it has been helped by its relocation. Its history of different sitings – outside a power station in the shadow of the 'Three Sisters', alongside a statutory corporation, facing a levelled mine – neatly captures fifty years of coalfield history. And although the sculpture has now come to rest, poised in time between the energy production of the past and the future, and in space

between the vacant ground of a sealed pithead and the regenerated dross of a departed industry, it is acquiring new meanings.

Most striking from a distance is its exposure; an isolated figure in a flat, windswept landscape under a big open sky, the antithesis of the airless, claustrophobic setting of a miner's daily work. Orwell, describing the physique of the fillers – 'wide shoulders tapering to slender supple waists… small pronounced buttocks and sinewy thighs' – wrote that they looked 'as though they were made of iron'. Viewed from the roundabout, 'The Waiting Miner' perfectly matches this description but from up close the torso of the figure, though wirily muscled, is also hollow-chested. This, together with its noticeably glazed, almost sightless eyes, suggests some of the diseases caused by mining, such as silicosis, pneumoconiosis, and nystagmus. The posture is stoical rather than heroic, patient and resilient, neither defiant nor defeated but marking something that has passed and waiting for something new to arrive. The obvious reference of the sculpture's title is to someone waiting for the cage to take him down into the pit, but it has acquired a new meaning as well, of waiting for a future that, although recently postponed, will nevertheless soon arrive. The disproportion between the plinth and its figure, the massive, raised base like a giant slab of coal contrasting with the sinewy physique of the crouched miner, also expresses something more timeless, the mutability of individual and social existence compared with the bedrock on which we live.

The history of the mining industry is now usually told as a story of struggle, defeat and extinction. Feature films and documentaries of its decline and fall are typically a ghostly kind of storytelling written from the perspective of a post-industrial world in which the mining landscapes of the past have been grassed over and forgotten. Bill Morrison's *The Miners' Hymns*, made in 2011, captures this mood and point of view most hauntingly. It opens and closes with an aerial sequence in which the camera moves slowly over the suburbs and business parks which have replaced the pitheads and spoil tips of the now vanished mines of the Durham coalfield while a voice intones the names of its lost pits: Ryhope Colliery, Silksworth Colliery, Hylton Colliery, Monkwearmouth Colliery. These shots are in bright colour, in contrast to the rest of the film – footage from every decade in the twentieth century of the region's mining past – which is in black and white. A depopulated and anomic present is set against a past that is teeming with people, memories and a now extinct culture: the Durham Miners' Gala, or 'The Big Meeting' as it was always known locally; men gathering coal from the shore, sieving it as they prospect for black gold; children playing cricket in the street and sliding down a spoil tip. This archival footage was always in black and white, of course, but its contrast with the garish colour of the world today gives it a spectral quality in which the miners are like shades, souls of the dead. This is reinforced by having all the footage from the past in slow motion, creating a phantasmal world of memories, glimpses and echoes. Jóhann Jóhannsson's score, the film's only other soundtrack, works to the same end. Victorian hymn tunes were part of the repertoire of colliery brass bands, and these are reworked to provide a solemn music that culminates in the apotheosis of miners and their lodge banners arriving at Durham Cathedral.

The Miners' Hymns is very moving but in an almost entirely melancholic and elegiac way. It memorialises a way of life seen as wholly vanished and which struggles even to be remembered, a narrative of obsolescence. Betrayal and extinction are certainly part of the story, but these themes resonate so powerfully that any possibility of salvage, adaptation or new beginnings is overwhelmed.

Unlike the Durham coalfield of Morrison's film, the mines and communities of the Kent coast have never entirely vanished from sight. Snowdown, left abandoned since its closure when the site reverted to its pre-nationalisation owner, a local landowner to whom the Coal Authority still pays rent, is a ruin, its buildings crumbling, weeds and undergrowth spreading across the site. But it is still there, a derelict reminder of a history that has refused to disappear. Tilmanstone has gone but its Colliery Heritage Group has established a well-stocked museum and archive of its past. Aylesham remains a pit village and Mill Hill a mining community, even though there has been no extraction of coal in the area for more than thirty years. And most conspicuously, 'The Waiting Miner' not only helps preserve a memory of mining in the area but expresses a future for the former East Kent coalfield, not one that has emerged phoenix-like from extinguished pits but rather from the efforts of former miners and their communities, in keeping with the rhythms of loss and adaptation that characterise the social history and geography of Kent's seaboard. There it sits, above ground, quietly insisting on being noticed.

There was another, more specific way in which the Kent coalfield reflected the history of the exposed and vulnerable topography of this coastline, open to the world with its ceaseless flow of arrival and departure. Those who came to work there were not just internal migrants seeking employment and a fresh start but also immigrants from many other countries: Australia, Belgium, Canada, Germany, India, Ireland, Poland, Lithuania and South Africa, for example. There was the Lithuanian Lukosevicius family who settled in Aylesham in the 1930s and later changed their name to Lucas; Philip Franks, who came to Betteshanger via the Kitchener Camp and whose mother was gassed at Auschwitz; Josef Maslak, who worked with the Polish resistance movement, was imprisoned by the Germans and came to Kent after the Second World War where he worked at Chislet until its closure in 1969 and then Snowdown; many more from other places who contributed to the rich underground life of Kent's coastal border.

In exploring the pull of underground worlds, I'd fallen under the spell described by my fellow volunteer at Fan Bay. When now I see a hole in the ground, I want to go down it. Subterranean spaces reveal the ground beneath our feet to be brittle or porous, less solid than we assume. I'd come to identify the below-ground world with the past, seeing it as a place of clues to, and stories of, lost worlds. Once underground, the loss of one's bearings is both stimulating and disorienting. There is a sense of freedom, of being beyond reach of the official above-ground world, even though that official world has itself often retreated underground, closing off its spaces to all but the authorised. But caves, tunnels and mines are also places of fear and danger: of the dark, of being trapped, of the tenebrous depths that a journey inwards as well as downwards can reach, of ghostly pasts, of death.

BORDERS

12

ASYLUM AND DETENTION

‡

I'm sitting outside the door of court seven in the Immigration and Asylum Chamber, Taylor House, Rosebery Avenue, London. The door has a narrow window through which I have a cropped view of the proceedings inside. The judge is out of sight; I have a back view of the barrister for the applicant; the Home Office solicitor is directly in my line of vision; and I can see K's face on a large, fuzzy screen high up in the back corner of the court. This is the closest that he will get to the proceedings. K is in a cell in the Dover Immigration Removal Centre from where he is witnessing the hearing that will decide his immediate future. He has been detained in the removal centre for more than two years now and today he is making yet another application to be released on bail while his case for asylum is decided.

I have travelled here from Deal to stand as K's surety. This is voluntary work I undertake for Kent Refugee Help, an organisation particularly concerned with detained asylum seekers. If bail is granted, I will be called into the courtroom, questioned about my opinion of his character and asked to put up a sum (normally £500) that I will forfeit if K breaks the conditions of his bail. I've attended these hearings before but I'm never quite sure when I'll be called into the court. Sometimes I'm allowed in from the beginning; sometimes it's only after a decision has been reached. Different judges lay down different rules. Many of the procedures in this court seem arbitrary.

K is physically excluded from his own hearing. In a criminal court, the accused is present at their trial, visible to judge and jury as still a free person. Conviction and imprisonment involve a radical change of status. K, however, though charged with no crime, is already a prisoner in a cell.

Rejecting his application merely involves leaving him where he is. Freeing him on bail is the radical step. K's face on the screen, from where I can see it, projects badly. He's a good-looking young man, softly spoken, an attentive and intelligent listener, but none of this comes across. His face looks oddly squashed, like the TV image of a convicted or wanted person, and the close focus of the camera cuts out his body language. K has told me that the sound quality in the detention centre is so bad it is difficult to hear what is going on, another form of exclusion from these proceedings. He is dehumanised by his bodily absence from the court, rendered an exhibit for the Home Office, visual confirmation of the endlessly recycled phrases of its solicitor: 'a poor immigration history', 'cannot be trusted to keep the terms of his bail', 'likely to abscond', 'refuses to cooperate' and so on. These phrases have been repeated application after application, like an answerphone on permanent loop.

I'd been visiting K in Dover for more than a year where he is being held in the Citadel, a moated keep at the very top of Dover's Western Heights, the stronghold at the centre of those fortifications that so infuriated William Cobbett. Unlike Dover Castle, which dominates the town, the Citadel and its surrounding fortifications are hidden away out of sight of the thousands who daily pass through Dover. Its walls, once intended as the last redoubt for those defending England's coast, now hold those whose presence is felt to threaten the nation's stability and identity. The moat protecting those walls is at least fifteen metres deep and the same across. The Citadel represents fortress Britain, defending the White Cliffs into which it has been dug from the latest imagined threat of invasion.

The signpost to the immigration removal centre at the bottom of Western Heights has a long-stay parking sign alongside it. Long-stay indeed. The road climbs steeply up past the fortifications that girdle the Heights. On visiting days, friends, partners, wives and children of detainees trudge their way up the hill past the Drop Redoubt and the foundations of a Knights Templars church towards the forbidding entrance of the Citadel. Near the top, the view suddenly opens out, harbour and docks at your feet, Shakespeare Cliff banking steeply to the south, the Alkham Valley at your back. Shipping dots the Channel. All this is out of sight of those held inside the Citadel where the cold and the wind are the only reminders of its exposed position. On every visit, no matter the season, I think of Tom Waits' line, 'colder than a well-digger's arse'.

Entering the Citadel involves passing through a series of locked chambers. First, I report to the visitors' shed where my details are taken and my possessions locked away. Pen and paper are not allowed inside, a prohibition on writing things down that extends through the whole system to the bail court, which is not a court of full record. The shed is chilly. A few toys have been provided for the small children who come with their mothers. There's a Van Gogh print on the wall, from his time in Arles when he cut off his ear. It's a painting of a small bedroom, cropped to give an enclosed, claustrophobic, cell-like feel. I guess the choice of picture is random. Cornfields wouldn't play any better in this place.

After a long wait, I'm sent across the bridge to the massive, frowning gates of the Citadel where I press the bell and wait some more. Inside I'm frisked; a sniffer dog checks me for drugs; and I'm given a wristband for identification. More waiting and then I'm taken across a courtyard to the locked door of the meeting area where the number on my wristband is recorded and I'm photographed. Some while ago, a visitor swapped places with a detainee who then walked out unchallenged, and the photograph is to prevent this happening again. I wonder who would have agreed to such an exchange. Finally, I report to a desk in the corner of the visiting area – the fourth check-in I've gone through – name the detainee I've come to visit and wait for him to be brought to me. It can take more than an hour after first reporting to the visitors' shed before reaching the person I've come to visit.

When Anne Owers was Chief Inspector of Prisons, she criticised 'the austere environment' of the visiting area at Dover and found that 'anti-bullying procedures were poorly understood and implemented'. This no doubt explains the plastic flowers in wall vases high up around the visiting area and a plethora of racial equality and anti-bullying policy statements. But this place could not be anything other than austere and depressing, and in fact the prison staff are friendly enough, unlike those, as several detainees have told me, in centres run by the private security firms such as G4S and Clearsprings. The separation of detainees from their wives, partners, children and friends extends even to the visiting area. The tables and chairs are fixed to the floor at odd angles to each other, making physical contact difficult. Detainees are not allowed to move around during visits. Embracing and kissing, even touching, is a complicated manoeuvre. On one of my visits, a detainee and a young woman were leaning forward to hold each other's hands as they sat and talked. "No touching!" one of the guards shouted.

The first detainee I ever visited was Z, an Iranian. He'd been picked up on a demonstration in Tehran after the disputed re-election of Ahmadinejad in 2009, detained, assaulted and released. When a warrant was put out for his rearrest, he went into hiding, escaped into Turkey and from there to Greece. Arriving in Dover with illegal papers, he was arrested, convicted of attempting to enter the country illegally and spent a number of months in several prisons in Kent. On completing his sentence, he was taken to Dover Immigration Removal Centre where he'd been detained for about a year when I first visited him.

He was in that state I soon came to find horribly familiar: anxious, depressed, embarrassingly grateful for any kind of support, indeed rather confused at being treated like a human being again, and lost in the labyrinthine intricacies of the immigration and asylum system. Before he could apply to be released on bail, he needed a bail address where he would be sent if his application was successful. Such accommodation is limited and there is always a wait. Once the applicant is given an address, it must be taken up within a fortnight, otherwise it is lost and he is back where he started. When this had previously happened to Z, the two-week window having closed before a hearing had been fixed, he'd cut his wrists. Next time around, the bail hearing was scheduled before the expiry of the address and so I made my first trip to the Immigration and Asylum Chamber.

Although a small number of law firms offer sympathetic and informed representation without charge, the supply is very limited and Z hadn't managed to secure one of these firms. Instead, like most bail applicants, he'd been allocated a solicitor from one of the approved law firms with Home Office contracts. This is seen as bottom-of-the-pile work, easy money for just turning up and going through the motions. On this occasion, Z's solicitor hadn't even bothered to turn up.

Because this left Z unrepresented, I was allowed into the court from the beginning of the hearing to speak more generally on his behalf than a bail surety would normally be allowed to. Although I knew the details of Z's case, my understanding of immigration law and its procedures was negligible. I'd been told that new evidence not previously available to the court was central to the application, so I emphasised his suicide attempt as evidence of the strain of long-term detention and drew attention to the warrant for Z's arrest in Tehran which had recently been obtained from his family. The absence of this at a previous hearing had been used by the Home Office to cast doubt on the plausibility of Z's claim of political persecution. I also pointed out that as Iran was refusing to take back people deported from the United Kingdom, Z was caught in limbo between a country that wanted to deport him and one that refused to take him. Unless granted bail, he would be stuck in the removal centre in Dover for the foreseeable future.

There was an adjournment while the Home Office solicitor took telephone instructions on this point. He returned to the court announcing with cheery insouciance, "Good news!" Several deportations had recently been made to Iran via the British Embassy in Paris and a more permanent arrangement was being negotiated with the Irish government. A number of countries from which asylum seekers have fled refuse to take them back. Even so, security companies will sometimes put them on a flight home knowing they'll be turned back on arrival. The companies have targets to meet, and these pointless and expensive flights add to the tally. There was no reason to trust the solicitor's reassurance.

My cross-examination by the Home Office solicitor was desultory, and when the judge delivered her ruling, I understood why. All the reasons I'd given for Z's release had become reasons for keeping him in detention. The suicide attempt showed that it would be unwise to release him; in the Citadel he could be kept under watch and provided with care. This was material for Joseph Heller. If he hadn't been in detention, he wouldn't have attempted suicide. The fact that Z's family in Tehran had been able to obtain the warrant for his arrest suggested the situation there was less dangerous than he claimed. The Home Office's assurance that deportation to Iran was possible was accepted without question and so any concern about protracted detention was set aside. Formal grounds for refusing Z's application then followed in a rush: there had been no appeal against the order for his deportation; no appeal against the rejection of his application for asylum; no fresh claim for asylum based on the recently obtained warrant. I knew nothing of all this and nor did Z, having had no legal advice as to what he should have done once his applications had been rejected. The judge concluded by saying there was a high risk Z would abscond if granted bail, that he wouldn't report if a removal order was confirmed and that bail was unnecessary when removal seemed reasonably imminent. Game, set and match to the Home Office whose solicitor had not even broken sweat.

Z's next bail hearing, three months later, was worse. I was kept out of the court until the judge announced his decision, and his summary rejection of the appeal meant I didn't even appear as a surety. Although Z had by now made a fresh claim for asylum and obtained a witness statement confirming torture and ill-treatment by the police in Tehran, his application was dismissed without any reference to this new evidence. Judge Malloy made it very clear that because Z had been convicted of trying to enter the country with illegal documents he couldn't be released on bail; 'danger of flight' was his repeated phrase. In fact, it is not an offence to travel with false documents if you are claiming asylum but the court solicitor at the time of Z's trial had not told him this, merely advising that he should plead guilty to try and minimise his sentence. Recently, after fresh legal advice, Z had applied to the Criminal Cases Review Commission to have his conviction examined, but Judge Malloy took no account of this either. An insecure criminal conviction would remain forever a reason for keeping him in detention. The judge reminded me of the Queen of Hearts in *Alice in Wonderland*. Don't worry about the evidence; let's get to the sentence. With little prospect of Z being returned to Iran, I left the court thinking he might be detained for the term of his natural life.

I was astonished, therefore, when just four days later, I learnt that Z was to be released, sent to a bail hostel in Oldham and tagged. Nothing had happened during these few days to change the facts of his case in any way. Indeed, he was now to be released without even the surety that I would have provided at the hearing. The judgement against Z's bail application had almost immediately been set aside by some Home Office official, concerned I suppose at the summary manner in which the application had been rejected and about a possible claim for unlawful detention.

I wrote to the president, First-tier Tribunal, Immigration and Asylum Chamber, Judge Michael Clements, asking him to explain these contradictory rulings. Here is his strange reply:

> *I cannot comment directly as to why bail was not granted by the judge nor why four days later the Home Office granted bail. I understand that the Immigration Officer is under a duty to consider whether bail should be granted on a continuous basis whereas an Immigration Judge would only consider it on the day of the hearing after an application.*

This was pure Gilbert and Sullivan. The distancing locution 'I understand' also made it seem as if the matter was really someone else's business. Judge Clements reminded me of one of the lesser Barnacles from the Circumlocution Office in Dickens' *Little Dorrit* (1857): 'You mustn't come into the place saying you want to know, you know'.

K had been held in the Dover detention centre for about eighteen months when I started visiting him and became his bail surety. He'd come to England in 2004, at the age of fourteen, with the help of a people smuggler and on a false passport. K is a Sikh, from a poor family in the Punjab. His family had arranged his journey to England after ethnic conflict had caused them to flee from their home near Amritsar and he had subsequently lost all contact with them. He worked illegally on building sites around London, had been removed to Belgium but managed to re-enter England and resume his former work. A dispute with his gang master over withheld wages resulted in a fight and K being charged with affray. As with Z, the duty solicitor at his hearing advised him to plead guilty and he was convicted without his defence being heard by the court. He served ten months in Wormwood Scrubs and on release was transferred directly to the Dover centre, and a deportation order was served.

K was in a relationship with a young Muslim woman who had a daughter, not his. They wished to marry and set up a home, and K planned to adopt the daughter. For religious reasons, his girlfriend's family was violently opposed to their relationship. She had suffered physical assault from her brothers because of it and was afraid to appear as a witness at any of K's appeal hearings. When it became clear that she wouldn't therefore be attending his appeal against deportation, his solicitors had closed the file.

When I first attended Taylor House as a bail surety for K, the Home Office solicitor and the judge between them had flung out his application. Confusion over the date of an earlier deportation hearing was described as a refusal on K's part to attend the court, confirming his 'poor immigration history' and his 'evasive', 'unreliable' and 'uncooperative' character. The loss of contact with his family and his uncertainty as to whether they were even still alive was interpreted as a wilful refusal to provide an address to which he could be deported. Whatever K said to explain or defend himself was turned against him as yet further evidence of his 'non-compliance', a word that echoes through Taylor House and an unanswerable charge when every attempt to counter it is taken as confirming the accusation.

The poet and academic David Herd, who has also attended this court as a surety, captures the branding and the double-binds at the close of his poem 'The hearing', in his collection *All Just* (2012). The poet sits at the back of the court listening to 'the man from the Home Office' cross-examining the appellant: 'Was it Dr Kumara / Who helped you leave your country / Or as you subsequently reported Mr Souma?'. Any confusion over names or dates or places is taken as a lie or an evasion. The poem ends:

> *And nobody present thinks to ask if you were scared*
> Sans papiers *pitched forward into shadow*
> Condamnee pour jamais a l'oubli de tous,
> *You saying, "Mister David it's all happening again."*

A month or so later, K was taken into hospital suffering acute chest pains. A heart attack was suspected but the hospital diagnosed anxiety and he was put on medication. At his next bail hearing, soon after his third Christmas in detention, this episode was presented to the court as new grounds for K's release on bail. The Home Office solicitor, however, argued that the medical report drew no connection between K's detention and his condition. Many people not in detention also suffer from anxiety, he said. This was an old defence, of course, like asking Betteshanger miners to prove that their lung conditions were caused by their job, or Medway naval dockyard workers that their cancers were the result of working around asbestos. The *Guardian* had recently reported that bonus incentives in the form of gift vouchers were being offered to Home Office officials with a strong record of successfully blocking immigration appeals. This solicitor had an eye on next Christmas.

The judge, though, was clearly anxious about the length of K's detention. Home Office guidelines state that prolonged detention should occur only when release would result in a significant threat to the community. Today K was represented by a very able young barrister, the first of any of the hearings I'd attended at which the applicant had been adequately represented. He pointed out that the Home Office had never suggested K represented any such threat. He appealed to the general principle of public law that published policy must be complied with in the absence of good reason not to do so. Failure to comply in this case would constitute unlawful detention. I wanted to add that K was now the second-longest detainee among the more than three hundred held in the Citadel, a different kind of record from a criminal one.

The judge was surprisingly open about her discomfort. She acknowledged the unacceptable length of time K had been detained, and even accepted that the confusion which had led to his failure to attend the deportation appeal was understandable given his circumstances and the complexities of the system. But she was concerned about his immigration record and particularly the conviction for affray: "Were tribal weapons involved?" she wanted to know. The man from the Home Office had argued that in the circumstances a surety of £500 was wholly inadequate and she agreed with this. In the light of all the positive things I'd said about K's character and reliability, was I prepared to put up £5000? This was a big ask. I trusted K to comply with his bail restrictions, but I had no idea how he would react if all his appeals failed and G4S came to take him to the airport. If I were in his position, what would I do? I was very conscious of K's face on the screen over my left shoulder. He was desperate to be released. What is the price of someone's freedom even if it is restricted and provisional?

I could see that the judge was caught on a hook from which my agreement to the large surety would release her. In effect, I'd be doing the Home Office's wretched business for them. The barrister tried to get the sum reduced to £3000 but the judge wasn't inclined to bargain. So I agreed; the judge spoke sternly to K about the responsibility he now had to me; and he assured her that he was very thankful and would do nothing to jeopardise my surety. She then thanked me as profusely as K had. His gratitude was touching and a bit embarrassing; hers was sickening. I'd never fully understood before that a bail surety was a kind of unofficial officer of the court.

Two days later, on a bitter January afternoon (a perfect Tom Waits day), K walked out through the gates of the Citadel pulling a ramshackle suitcase holding everything he owned. It was mid-afternoon and he had until the end of the day to reach his bail accommodation in North Shields, Tyne and Wear. He'd been given rail vouchers to take him to Newcastle but no advice on how to get there, nor any money for the journey. I drove him to Dover Priory station, gave him some money and put him on a train to Charing Cross, wishing I had the power to dump the home secretary and the chief executive of G4S in Chandigarh with instructions they must reach Bangalore via Delhi before midnight. I drove home to Deal thinking it would be a miracle if he reached his bail address – Hutton House, North Shields – that day.

The following afternoon, I got a text from K: 'Hi Rod, I'm in Hutton House just half an hour ago I came in. They stand me on the road last night for 4 hours and today all day I was standing on the road in the cold'. I got the full story from his case worker at Kent Refugee Help. K had arrived at Hutton House by 10.30pm, but there was no one to meet him and let him in. He'd rung the case worker; she contacted the police; and he'd been put up overnight in a cell. Next morning the police had taken him to the reporting centre in North Shields where, under the terms of his bail, K must attend each Monday and Friday. There he was told by his case owner, an employee of G4S, that his bail accommodation had expired and that he'd be better off going back to the Punjab.

After further intervention from the case worker, K was finally tagged and admitted to his hostel almost twenty-four hours after his release from Dover. G4S had failed to meet K as had been arranged and were wrong about his accommodation having expired. He was distressed by his treatment and had obvious grounds for a complaint but was reluctant to do so for fear of making his life even more difficult. In this way, the Home Office and the security industry are protected from their abuse of the people they are responsible for.

The euphoria of release from detention is short-lived. I knew this from Z's experience of being bailed to Oldham, a place he had never heard of, where he knew nobody and where the only focus of his week was attending the reporting centre. I hadn't liked to tell him that Oldham had experienced race riots in 2001 and was where the leader of the British National Party had polled almost fifteen per cent of the vote in that year's general election.

K's texts to me over the next couple of weeks expressed his growing sense of the emptiness of life in a bail hostel: 'Hi rod, im fine how are u? U know its new area for me hopefully everything will be OK slowly slowly'; 'Hi Rod, im OK how r u? I'm feeling boring no friends no tv here strange area'; 'Hi Rod im fine. How are u. And X [his girlfriend] still in london because we don't have money. we still in hard time because im too far away, thank you'.

For more than two years in the Dover Immigration Removal Centre, K's life had focused on being freed to live with his girlfriend. But from the moment of his release from the rudimentary comforts of the Citadel – warmth, TV and bad food – to the exposure of his life outside and the realisation that living with his girlfriend was as far away as ever, K was faced with a whole new set of difficulties that bail is designed to reinforce. He had very little money, an Azure card – a voucher for basic needs – to be used at a few designated shops, no freedom of movement, no chance of work, no friends. To be out on bail, he discovered, was a bleak and constricted kind of release, a continuation of the social and psychological isolation of life inside the Citadel.

Once out on bail, the asylum seeker is confronted by the hostile environment which successive governments have created to persuade them to 'voluntarily' return to their country of origin. For example, around the time of K's release from the Citadel, G4S painted the front doors of the houses in Middlesbrough where they had placed asylum seekers released on bail, red. Thus identified and stigmatised, the inhabitants were subjected to hostile knocking on their red doors, stones being thrown at their windows and verbal abuse. I doubt if G4S realised the biblical significance of the red door, but in the Book of Exodus a door marked with the blood of a lamb signalled a Jewish household, one to be spared from God's wrath. G4S managed to reverse the story, a safe house becoming a target of wrath. Z in Blackburn and K in North Shields experienced similar kinds of abuse. Not all prisons are locked.

The novelist and Nobel Prize winner Abdulrazak Gurnah, himself a refugee from Zanzibar in the late 1960s, described Dover in the 1990s as 'an open detention centre', an apparent oxymoron but an accurate description of the living conditions of many refugees, whatever their formal status, struggling to survive in the United Kingdom today. Pawel Pawlikowski's film *Last Resort* (2000), which imagines the whole town of Margate as a detention centre, further blurs the distinction between being detained, being out on bail or just existing as an asylum seeker.

Pawlikowski, who came to Britain from Poland in his teens, recreates Margate, Stonehaven in the film, as a place of high fences, armed guards, surveillance cameras and dogs. We have glimpses of the bleak shoreline of Thanet stretching from the towers of Reculver to the now vanished cooling towers of Richborough, and the skies that Turner painted also engage Pawlikowski's camera, but most of the film is within sight of Margate's stone pier where the Turner Contemporary gallery now stands.

Tanya, a Russian, and her ten-year-old son, Artiom, arrive at Gatwick expecting to be met by her 'fiancé', but he fails to show. Lacking any means of support, Tanya panics and claims political asylum. She and Artiom are then taken to Stonehaven. It is some while before Tanya fully understands her situation. There is no way out of the town. The station has been closed, and there are no cars or buses in Pawlikowski's Margate. Her claim for asylum will take more than a year to process, but withdrawing her claim and getting released to go home will take at least as long. Like many refugees, she is trapped in a non-place without civil or legal rights or adequate support.

Another of the detainees I stood surety for experienced something very similar. H, an Algerian, who had lived in Britain illegally for more than a decade, found the hassles of such a life were no longer worth it and decided to return home. But being *sans papiers*, he was detained in the Citadel for a year while the Algerian Embassy, a notoriously slow office, failed to deal with his request for a passport. After several unsuccessful bail applications, the absurdity of detaining someone whose only wish was to leave the country finally secured him release to a bail hostel in Plymouth where, two years later, when we were last in touch, he remained.

There is nothing overtly repressive about detention in Pawlikowski's film. The officials are indifferent but not unpleasant. The guards leave the detainees alone unless they try and escape. Nevertheless, we see Tanya and Artiom's world shrink to the size of the centre of Margate. Life beyond is out of reach and their existence is reduced to mere survival and the desire to escape. The derelict seafront and the sombre tones of the sea and the sky reflect their situation.

At the time of the film's making, Margate's empty guest houses were being used by London councils as a dumping ground for their homeless. These internal migrants joined refugees and displaced people who had washed up in Margate: Czech and Slovak Roma, Kurds, others from Bosnia, Kosovo, the Congo, Belarus, Turkey, Egypt, Afghanistan, Albania. Margate, with the highest unemployment rate in the south-east, became known as 'Dole-on-Sea' or 'the cat flap at the end of the A28'. If Dover had been the point of entry for those seeking asylum, Margate was the end of the line. Pawlikowski folds all this into his film. Refugees walking the streets were recruited for crowd scenes. We see them lining up to give blood as a way of earning some money. A group of Turkish musicians appears several times, in their crowded, run-down flat, on the seafront after dark. The kids he cast as the small gang Artiom hangs out with were known locally for abusing and spitting at asylum seekers. Appearing in the film meant pocket money for all these people.

Dreamland is also incorporated into Pawlikowski's vision of Margate as a detention centre. When *Last Resort* was filmed, Dreamland had closed and many of its attractions, including its famous big wheel, had been sold off. Its stripped-down semi-abandoned state perfectly catches the in between situation of Tanya, Artiom and the other inmates of Stonehaven. Emptied of its pleasures, Dreamland has become a place of lost or empty dreams. Tanya looks down on the dilapidated site from the window of her bleak flat near the top of Arlington House, Margate's monument to the high-rise brutalism of the 1950s, where she and her son have been put. Artiom and his friends swig vodka on one of the immobilised merry-go-rounds. Dreamland has decayed and its lights have gone out. Migration has ended in a wasteland of hope, its beached characters stranded among the wreckage.

Stonehaven is an amalgamation of a detention centre with a bail hostel – a kind of open-air prison – and stands for the plight of most asylum seekers once they have crossed the Channel in search of sanctuary. Tanya and Artiom – like Z in Oldham, K in North Shields and H in Plymouth – have more freedom and less hour-by-hour interference in their lives than behind the locked doors of a detention centre. But they all remain trapped, their life on hold, their movement circumscribed, under the control of those security companies who have benefitted so profitably from the asylum system. Indeed, from the point of view of the asylum seeker, the whole country can seem like a detention centre or open prison.

The Napier Barracks near Folkestone is a recent version of Stonehaven. Once part of Shorncliffe Army Camp, established like so many fortifications along this coastline during the Napoleonic Wars, it was hastily reopened in September 2020 as temporary accommodation for single adult males whose asylum claims were under consideration. The use of a military base as a refugee camp was a new step for this country. Rather than providing the 'initial accommodation' that asylum seekers are entitled to, they were detained in what was effectively a prison camp and a holding compound for deportations.

Covid quickly took hold in the camp. Public Health England had warned the Home Office that the Napier Barracks was unsuitable for housing refugees, and the absence of any means of social distancing soon resulted in almost two hundred cases, about half the number of those held there. The home secretary, Priti Patel, blamed this on the detainees for 'mingling' in their fourteen-bed dormitories. Far from mingling, they had been hanging sheets between their beds in a desperate but hopeless attempt to stem the spread of infection in their overcrowded and insanitary dormitories. In protest at these conditions, one of the buildings was set on fire, an action described by Patel as 'deeply offensive to the taxpayers of the country'.

Six of the detainees then brought a case against the Home Office, defying threats from security guards at the barracks that this would jeopardise their applications for asylum. Similar threats had been made to detainees who had spoken to the media about the conditions in which they were being held. The High Court ruled that the detention of asylum seekers at Napier Barracks was unlawful and criticised its 'detention-like setting'. Asylum seekers should be living voluntarily while their applications were considered but instead, they were being held 'in a site enclosed by a perimeter fence topped with barbed wire, access to which is through padlocked gates guarded by uniformed security personnel'.

The transfer of asylum seekers to Napier Barracks was then briefly suspended but resumed within a couple of months and a five-year deal with Clearsprings, the security firm running the camp, was announced by the Home Office: 'While pressure on the asylum system remains', it said, 'we will use Napier Barracks to ensure we meet our statutory duty. Asylum seekers are staying in safe accommodation, where they receive three nutritious meals a day paid for by the British taxpayer'. Put this against the reality of life inside Napier Barracks as described by one of the inmates: 'I worry about the people who spend the whole day in bed. They wake up to eat and then go back to bed. They do not speak to anyone, they cannot speak English nor do they have a phone or anything. There are people like this in every building'.

It is not only opponents of detention who are unhappy with the continuing use of the Napier Barracks. The Dover MP Natalie Elphicke, notorious for suggesting that if Marcus Rashford had concentrated on football rather than campaigning for children to be adequately fed, England might have won the European Championship, insisted that it 'sends entirely the wrong message… All efforts should be made on bringing the small boats crisis to an end, not accepting it will continue for many years to come. No one should be put anywhere near Dover, as it risks creating another migrant magnet'.

Soon after the renewal of Clearsprings' contract, I visited the barracks. The steep and winding drive up Military Road to the old Shorncliffe camp reminded me of the climb up to the Citadel, both places of detention situated on two of the highest and most exposed points along the East Kent coastline, facing outwards towards the worlds from which the asylum seekers they hold have fled. Nothing could better express their precarious and unwanted state. Napier Barracks, to repeat Defoe's phrase, is another 'just reproach to all the land'.

It was a bright, windy Sunday morning, and local residents were out with their dogs, strolling along the footpath across the road from the high wire fence and heavily padlocked gates of the barracks. The barbed wire topping the fence had gone but been replaced by a heavy screen strung along the fence, obscuring the view both in and out. I peered through a narrow gap in the screen at Gate 14. A young man in a Manchester United shirt – No 10, Rashford on its back – was standing in the middle of an exercise yard, one foot on a worn-looking football, staring at the ground, motionless. I moved on along the fence past the fire-gutted dormitory which the home secretary had thought such an insult to the British taxpayer. Another young man was standing on the steps of the dormitory alongside it. He called out 'good morning'; I returned his greeting; and we gave each other a wave as I passed on down the road. I should have stopped to talk to him. Why didn't I? I think I was trying to avoid the uncomfortable tangle of feelings I remembered when visiting detainees in the Citadel.

The locals out walking their dogs paid the camp no attention, but several stared at me suspiciously, clearly wondering what I was doing walking on the wrong side of the road with a notebook and pen in my hand. Their dogs seemed rather interested in me too. Folkestone is renowned for the hospitality it extended to Belgians fleeing from German invasion during the First World War. On one single day in August 1914, around sixteen thousand Belgian refugees, double the population of the town, were welcomed at Folkestone Harbour. This spirit of greeting had not endured.

Clearsprings, an Essex-based company, has Home Office contracts for asylum accommodation across the south of England and Wales, including the Penally camp in Pembrokeshire, the other military base currently being used to hold asylum seekers. Clearsprings is a wonderfully inappropriate name, suggesting something refreshing and life-enhancing, sanitising the reality of the conditions in which those it houses are kept. Flats the company provides in Uxbridge and South Ruislip (Boris Johnson's former constituency) were found to be infested with rodents and cockroaches, with water leaking from walls and ceilings and the electricity switched off at night to save money. An eighteen-year-old Sudanese asylum seeker housed there would cover his bed with plastic bags when it rained and sleep on the floor under it. When charity workers raised these matters with Clearsprings, the manager replied 'lol'.

Clearsprings is one of three outsourcing companies – Serco and the Mears Group are the other two – with Home Office contracts to provide asylum accommodation across the United Kingdom. G4S lost its contract following the exposure of its abusive treatment of detainees at the Brook House Immigration Removal Centre at Gatwick. The outsourcing system which allows these companies to put profits before the living conditions of those they're responsible for goes back to the end of the 1990s when the Labour government introduced a dispersal policy that sends asylum seekers to the poorer regions of the United Kingdom. As a consequence, the contractor companies have crammed their charges into cheap, run-down accommodation in areas with already strained and under-resourced local services, overwhelmingly in the north of England. The government has plans to open two new detention centres for asylum seekers waiting to be deported to Rwanda. More money for the outsourcers.

Very rarely do companies such as Clearsprings own the properties in which the asylum seekers are housed. They are middlemen between the Home Office and local slum landlords or large property companies, outsourcing the day-to-day management of these properties and thereby camouflaging or evading their responsibility for failings and abuse. Napier Barracks, for example, has been subcontracted by Clearsprings to Nationwide Accommodation Services; the running of its many London hostels and houses to Mylandlets; the Uxbridge flats to Cromwood. And the Bibby Stockholm barge, the floating hulk moored in Portland Point to detain asylum seekers and initially outsourced by the Home Office to the Australian firm Corporate Travel Management (another name beyond satire), has been further subcontracted to a Miami firm, Landry

& Kling. This pattern of multiple outsourcing oils the system under which asylum profiteers can flourish.

At the start of 2022, a processing centre for asylum seekers arriving in small boats at Dover was established at a disused airfield on Thanet: Manston, a former military base. It was designed to hold a maximum of 1,600 people for no more than twenty-four hours while security and identity checks were made. By autumn of that year, the camp was holding up to four thousand people for periods of more than a month in some cases. When the chief inspector of borders and immigration, David Neal, visited the centre, he was 'left speechless by the wretched conditions he saw'. Those detained there were held behind barbed wire, had no idea where they were, were referred to by the numbered wristbands they wore rather than by their names and were prevented from calling their families to let them know they had safely reached the UK. 130 people, children among them, were sharing a single large tent. Cases of diphtheria were discovered, a bacterial disease particularly dangerous for young children – the Victorians called it the 'strangling angel' – which spreads in crowded and unhygienic conditions. Some of the security staff were untrained and without Home Office accreditation.

The outsourcing company running the centre is Mitie (Management Incentive Through Investment Equity). Described on its website as 'the UK's leading facilities management company', it 'helps partners achieve strategic goals and create amazing environments', both of which it has certainly achieved at Manston. David Neal was so amazed by the environment it left him speechless, and the Home Office's strategic aim of ensuring a hostile environment for refugees has been met to the full. Mitie's headquarters, by contrast, are commodiously based in The Shard.

Whether held in processing or detention centres, bail hostels, houses or any other form of confined accommodation while they wait for the Home Office to decide on their future, life for asylum seekers living under the regime of 'accommodation and support' provided by these outsourcing companies on behalf of the Home Office is fundamentally the same, a core part of the hostile environment created by successive Labour and Conservative governments. And the issue of detention is set to become much worse with the introduction of the draconian 2023 Illegal Migration Bill. The plan to detain asylum seekers immediately on arrival and hold them in detention camps as they await speedy removal to Rwanda or Neverland will result in mass detention on a scale never before seen in this country.

13

MIGRATION AND STIGMA

‡

Primo Levi, in his account of surviving Auschwitz, *If This Is a Man* (1947), describes the inherent cultural tendency to stigmatise others:

> *Many people – many nations – can find themselves holding, more or less wittingly, that "every stranger is an enemy." For the most part this conviction lies deep down like some latent infection; it betrays itself only in random disconnected acts, and does not lie at the base of a system of reason. But when this does come about... then, at the end of the chain, there is the Laager.*

The laager – a defensive encampment formed by a circle of wagons – graphically captures the United Kingdom's response to the displacement crisis of recent decades.

In Dover in 1997, an influx of Czech and Slovak Roma – triggered, it was said, by a programme on Czech television, *Gypsies in Heaven*, highlighting the welfare benefits and dole money to be enjoyed in Britain – prompted an ugly local press campaign against Roma asylum seekers. 'Human sewage' was the phrase that provoked even the spineless Press Council to censure the editor of the local paper, a publication already notorious for a widely circulated story under the headline 'Asylum Seekers Ate My Donkey'. The National Front organised rallies in Dover. The sound of its members and supporters marching through the town chanting 'The National Front is a white man's front' and singing 'Rule Britannia' can be heard in a radio documentary made at the time by the Nobel laureate, Abdulrazak Gurnah. Gurnah knew the territory he was investigating. He had arrived in Dover as a refugee from civil war in Zanzibar many years earlier, had once taught in a school in the town and, in a succession of novels from *Pilgrim's Way* (1988) to *By the Sea* (2001) and *The Last Gift* (2011), has explored the uncomfortable margins that immigrants – 'strangers' – inhabit.

His radio documentary, *Scenes from Provincial Life*, included reflections on his own arrival. This had been an experience of loss – of family, place and identity – rather than grateful or greedy satisfaction at enjoying the benefits of British life. A child of empire himself, brought up on Britain's reputation for benign tolerance, he found himself treated like an unwelcome stranger. The Czech and Slovak Roma in 1997 underwent a more intense and physical form of the same experience, with verbal abuse, spitting, assaults and hostility to their children at local schools. As so often, this resentment was fed by other local discontents, a feeling of neglect and abandonment common to many eastern coastal towns exacerbated by the way that central government had shifted the cost of dealing with asylum seekers to cash-strapped local authorities.

Resentment, though, did not always preclude sympathy. One of those Gurnah spoke to, a woman who provided bed-and-breakfast accommodation for Roma, was receptive to the stories of suffering, people having their houses burnt down for example, she had heard. Individual accounts of persecution or loss can induce sympathy that, temporarily at least, overcomes a general prejudice against a whole group, challenging the instinct to blame the refugee for difficulties that local people face in their own lives.

Nevertheless, the bed-and-breakfast owner wondered, *Are they all genuine?* The figure of 'the economic migrant' spinning a hard-luck tale to justify their desire for a better life is a let-out for those who might otherwise be moved to generosity. So-called economic migrancy is usually a result of discrimination, persecution or displacement. The mild scepticism of the bed-and-breakfast woman was a softer version of the attitude of the bail judges that Z and K had come before. Prove to me that your story is true, they ask. Prove to me that you are not just trying to take advantage of our goodwill. But how do you prove that your story, which starts thousands of miles away, which lacks producible witnesses and which, like any story, is bound to have loose ends, is true? If there are apparent gaps or holes in your narrative, then everything you say becomes suspect.

Since the turn of this century, different 'risk groups' have been identified, stereotyped and stigmatised at different times and in different ways: working too hard and taking our jobs in the case of the Polish; not working at all and taking our benefits in the case of Romanians. The target of immigration fears has been a shifting one. Tony Blair, during his second term as prime minister, 2001–2005, was particularly concerned about so-called 'illegal immigration' and 'asylum abuse' from beyond Europe. In 2003, Downing Street collaborated with *The Sun* on a special 'asylum week', a series of articles that began with a piece headlined 'Halt the Asylum Tide Now' and ended with then Home Secretary David Blunkett promising 'draconian measures' to clamp down on 'illegal immigration'. Two years later, Blair was reported as saying: 'The one thing that could lose me the next general election is immigration'. Like other politicians, he chose the White Cliffs of Dover for an election speech on immigration and crime. He did not see immigration from within the EU as a problem and, unlike most other member states, Britain imposed no transitional controls when ten new countries joined the EU in 2004.

Within a few years, the distinction between the inwards flow of new EU citizens and the migration of those from North Africa and the Middle East escaping their conflict-torn countries was becoming blurred amid a rising clamour about immigration in general. David Cameron's Conservative-Liberal Democrat coalition government, with Theresa May as Home Secretary driving immigration policy, announced the creation of 'a really hostile environment' for those living here illegally, an environment in which Nigel Farage and UKIP would flourish. Farage's 2013 conference speech, its verb-less ramblings reminiscent of the charlatan and trickster Alfred Jingle in Dickens' *The Pickwick Papers* (1837), seized upon May's 'hostile environment' policy: 'Ten thousand a week. Half a million a year. Five million economic migrants in 10 years coming to this country. Unprecedented. Never happened before'. He went on to blame all the shortages in public services on this list of fake statistics.

Soon the hostile environment policy was in full swing. Billboard vans with slogans telling 'illegal immigrants' to 'go home or face arrest' circled around six London boroughs with high immigrant populations. Immigration Enforcement vehicles were branded like police cars and adverts with the 'go home or face arrest' message were placed in minority ethnic newspapers. The wider scope of this strategy was to enlist landlords, employers, banks, doctors and teachers as immigration enforcement officers, checking the status of those of non-British nationality or, inevitably, British citizens of colour when they applied for a job, went to a surgery or hospital, tried to obtain a driving licence or otherwise went about their daily life. The Windrush scandal was an unforeseen consequence of this new catch-all regime. Immigration Acts of 2014 and 2016 gave statutory weight to these and many other hostile policies.

Labour's position on immigration floundered this way and that. Gordon Brown's pledge at his party's conference in 2007 to create 'British jobs for British workers' was a recognition that European immigration had proved heavier than anticipated and had become a political problem for Labour. Like many subsequent bold declarations of anti-European sentiment by British political leaders of both main parties, this was an empty boast. There was no basis in law under the terms of EU membership to discriminate in favour of British workers. The phrase 'hostile environment' was actually first used by a Labour home secretary, Alan Johnson. Labour's response to May's aggressive immigration policies was to drift with the tide; its sad red mug inscribed with the words 'Controls on Immigration' during the 2015 general election campaign, repeated on the even sadder 'EdStone' that marked the graveyard of Labour's election hopes, encapsulating its woeful response to UKIP and the hostile environment strategy.

The 2016 EU referendum ran together many different types of 'stranger' into a single image that had no direct relevance to the referendum itself but captured the feeling that Britain had been swamped by outsiders and was losing its distinct identity as an island nation. Farage's notorious 'Breaking Point' poster, in fact a photo of Syrian refugees entering Slovakia, exemplified this. Government policies, Labour's vacillations, the xenophobic clamour of the *Daily Express* and the *Daily Mail*, and the effect of years of austerity were folded together to legitimise public rancour against a generic 'illegal immigrant'.

I was caught up in one such display of this in the year of the referendum. Far-right groups headed by the National Front and Britain First gathered in Dover protesting against immigrants in general and asylum seekers in particular. Local pro-immigration groups called a counter-rally in response. I joined a crowd of three hundred or so in Market Square in the centre of the town to hear the speakers: Diane Abbott, Labour's Shadow Home Secretary, and someone from the Calais Jungle just across the water, among others. A similar number of far-right demonstrators were meanwhile assembling at Dover Priory station to march to the Eastern Docks carrying banners such as 'Save Our Country from Invasion'. Combat 18 T-shirts were prominent; there was even one blazoned with Enoch Powell's name. Many of them carried poles, bricks and other offensive weapons. Someone shouted 'I hope you get raped by a refugee' at the *New Statesman* reporter, India Bourke.

After the rally, fights broke out between the far-right demonstrators and a counter-group of masked and hooded anti-Fascists – North London Antifa, I was told – chanting 'Smash the fascists into the sea'. There were many arrests and convictions, mostly of far-right demonstrators. One of them caught on camera beating a photographer with a Union Jack flagpole and splintering his elbow was given seven years. As the list of convictions showed, the anti-immigrant demonstrators had come to Dover from all over Britain: Romford, Anglesey, Sheffield, Merseyside and Edinburgh, for example. There were several from Blackburn and South Shields, parts of the country where Z and K had been sent. Dover was once again the nation's symbolic stronghold, the point of entry to be defended, the place where those 'random disconnected acts' described by Primo Levi become a concerted resistance to all strangers.

In 2015, the number of people worldwide forced to flee their homes exceeded fifty million for the first time since the Second World War. Because these refugees came from outside Europe, Britain and other EU countries arrogated the terms 'problem' and 'crisis' for themselves, diverting attention from the fact that it was primarily and overwhelmingly a problem and a crisis for all those Syrians, Afghans, Iraqis, Eritreans, Somalis, Libyans and others who were desperately seeking safety. It was a comparatively small problem for Europe and a very small one indeed for Britain. The large majority of refugees, eighty-six per cent in fact, sheltered in less developed regions or countries. More than four million Syrian refugees, for example, were spread across Turkey, Lebanon, Jordan, Iraq and Egypt.

Since then, the number of refugees reaching Europe has significantly diminished, mainly as a result of agreements struck between the EU and some of its neighbouring countries, such as Turkey, whereby refugees are intercepted and detained in return for financial and other forms of aid. A similar deal was struck with Libya. As a consequence, although the number of refugees crossing the Mediterranean has fallen, the proportion of deaths among those attempting the crossing has increased. The Greek police have forcibly recruited detained asylum seekers to assist in violent and illegal 'pushback' operations. The Italian government has prosecuted crew of the rescue ship *Iuventa* for saving refugees attempting to cross the Mediterranean.

The British government has joined in these efforts to turn the Mediterranean and the Aegean into Europe's moat. HMS *Bulwark*, which had been rescuing refugees from the Mediterranean, was replaced with a survey ship, HMS *Enterprise*, whose sole task was to gather intelligence on 'migrant flows' and to prevent people smuggling vessels from leaving the North African coast. British troops joined NATO forces in the Aegean, turning back refugees trying to reach Europe from Turkey, a violation of both EU and international human rights law. In effect, Britain has extended its own border to the southern edge of Europe.

These measures to hold back unwanted migration merely shift the problems caused by human displacement and often exacerbate them. Refugees forced back to Libya find themselves trapped in detention centres where horrifying abuse, especially of women, is commonplace. As death tolls have fallen in the Mediterranean waters, they have risen in detention centres along its southern coast. Militias in these countries have long been involved in human trafficking and now, in the pay of Brussels, have no interest in providing safe crossing when they are rewarded for ensuring that migrants don't reach the opposite coast. Europe's curbing of migration in the Mediterranean has caused many refugees to attempt the treacherous Canary Islands route in flimsy boats unfit to handle fierce Atlantic currents. More than 4,400 died or disappeared trying to reach Spain this way in 2021, double the previous year's number and the highest since records began in 2007.

While the British government has assisted in keeping asylum seekers from reaching Europe, its overriding concern has been to prevent them from crossing the Channel. Those who make it as far as Calais are seen as especially undeserving because they have journeyed too far and come too close. Over the last twenty years, the right-wing press and politicians of all parties have created, to use Primo Levi's term, a 'system of reason' to explain why so many in this country struggle to find jobs or earn enough to live comfortably, why public services have declined, why the nation itself is not what it was. By systematic use of the refugee card, which is invariably also the race card, and by the cynical politicisation of asylum, successive governments have found a way to explain our ills and to justify the flagrant abuse of human rights that has increasingly characterised recent Conservative administrations.

This 'system of reason' has been broadcast and sustained by the recurring declaration of a crisis. When a number of small boats carrying asylum seekers came ashore or were intercepted off the Kent coast during the Christmas period of 2018, the right-wing press announced a 'migration crisis', and the home secretary, Sajid Javid, cutting short a holiday in South Africa, declared a 'major incident'. Local MPs joined the chorus, and Sajid Javid took the well-worn route to Dover where he described the recent arrivals as 'illegal immigrants' and promised he'd ensure those who crossed the Channel in this way were unsuccessful in their asylum claims. This was in clear breach of international law, and he was forced to withdraw his threat, but it fed and legitimised a rising clamour of anti-refugee feeling. A group of self-declared 'patriots' formed a vigilante organisation, South East Coastal Defence (SECD), to monitor the coast from Deal to Dungeness. Project Kraken, a multi-agency campaign involving the UK Border Force, the National Crime Agency, Kent Police and the Association of Chief Police Officers, distributed a striking poster of a black, hard-edged image of a lighthouse, rather like South Foreland in shape, its beam blazoning the words 'Seen Something Suspicious?', echoing public safety and vigilance posters from the Second World War. Dover, a portal for the displaced in two world wars, was now emphatically a point of no entry.

The 'Channel crisis' of 2021 was even more patently a manufactured one, explicitly designed in this case to assist the passage of the Nationality and Borders Bill. Although there had been a steady increase in the number of refugees crossing the Channel in small boats, overall asylum applications were falling. The figures for 2020 were twenty-four per cent down on the previous year, tightened security at the French ports having resulted in small boats replacing lorries as the readier way to cross the Channel. The reason for increasing numbers coming ashore on the Kent coast were well known to the Home Office but never acknowledged. Home Office delays in processing asylum claims and appeals against removal had caused a backlog of over one hundred thousand cases, which was also exploited to give the impression of a system overwhelmed by a recent flood of arrivals. The use of military camps like Napier Barracks and Penally as asylum holding centres made more visible and immediate the impression of a crisis. A new office of Clandestine Channel Threat Commander was created. Something had to be done.

The home secretary, Priti Patel, grabbed the headlines with a series of draconian proposals to keep asylum seekers at bay. The first of these was 'pushback'. The installation of giant wave machines in the Channel to force small boats packed with refugees back to the French coast was suggested. The impracticability, inanity and breathtaking indifference to human life of this makes me doubt it was ever a serious proposition, but it certainly heightened the sense of a crisis demanding a punitive response and prepared public opinion for the Nationality and Borders Bill as it began its passage through Westminster.

Instead of a wave machine, Patel then approved plans for UK Border Force to turn back refugee boats in the Channel, leaving it to the French coastguard to intercept them as they re-entered France's territorial waters. This was a flagrant contravention of international maritime law under which the safeguarding of human lives at sea has priority over all considerations of nationality, status and migration policy, as well as being in breach of international human rights law. Questions of law aside, the risk to life in turning back these flimsy boats is very high, especially in rough seas. The weather in the Channel, as I know from personal experience, can suddenly change. These vessels would be at risk of capsizing if a Border Force cutter approached at speed. Drownings have occurred in the Mediterranean when refugees have leapt into the water at the sight of Libyan coastguard vessels bearing down. The approach of a Border Force vessel would have a similar effect in the Channel.

Border Force staff were seen from Dover being trained in 'pushback' using jet skis, and the Nationality and Borders Act gave them legal protection if anyone drowned as a result of this strategy. The death of migrants, which is a scandal when traffickers can be blamed, apparently matters not at all if caused by Border Force, a view its employees and union, the Public and Commercial Services (PCS), certainly don't share. As one union official remarked: 'If somebody dies, it won't be Priti Patel taking the body out of the water'. The PCS challenged this practice and compelled the Home Office to abandon it but, renamed by Liz Truss as 'turnaround tactics', it remains a potential weapon in the Home Office armoury.

Patel's other big idea was the removal of asylum seekers to offshore processing centres. Long mooted – remote Pacific and Indian ocean islands, Albania and Ghana had previously been suggested – the deal struck with Rwanda early in 2020 has become the showpiece of the Home Office's answer to the so-called refugee crisis. This policy has the stated aim of deterring asylum seekers and destroying the traffickers' 'business model'. The traffickers, however, cut their prices and crammed more people into each boat. As a result, the number of asylum seekers crossing the Channel in the summer of 2022 hit record levels; the danger to life involved in making the crossing increased; and the traffickers have developed a new 'business model' as profitable as the former one.

Rwanda is a small and very poor country recovering from a dreadful civil war and with a widely criticised human rights record. It has almost twice the population density of the United Kingdom and is already hosting five times as many refugees per capita, approximately 140,000 people living in six refugee camps. It is also itself a source country for refugees escaping persecution and torture by the Kagame regime. Patel's deal, under which Britain has paid Rwanda £120 million for taking its cast-offs, treats those who have come to this country seeking sanctuary as mere commodities in which the terms of trade are like any other agreement between one of the world's ten richest countries and one of the twenty-five poorest.

Like most Home Office refugee policies, the arrangement with Rwanda is a mixture of cynicism and incompetence. Natasha Walter has described it with deadly accuracy as 'performative cruelty'. If deterrence was the real aim, it would require the deportation of many thousands of those claiming asylum in any one year. So far, the accommodation on offer is a seventy-two-bedroom hostel in Kigali: Hope House, a name beyond satire. And the whole policy is based on an obvious contradiction. On the one hand, the government has assured its many critics that Rwanda will offer asylum seekers the chance to rebuild their lives in a friendly and supportive environment. On the other, being sent there is so awful it will stop them from trying to reach the United Kingdom and will put the traffickers out of business.

In June 2023 the court of appeal ruled that the plan to pay Rwanda to take asylum seekers was illegal. Rwanda, it declared, was not 'a safe third country'. The government immediately announced its intention to appeal against this judgement. Whether successful or not, the appeal will enable a performative fight with the judciary, with the United Nations refugee agency, UNHCR (which testified in court to Rwanda's human rights abuse of refugees), and the European convention of human rights while positioning the Labour opposition where it doesn't really want to be in the run-up to next general election.

The 2023 Illegal Migration Bill (well named as most of its measures are clearly illegal) is the latest and most callous of the torrent of anti-refugee legislation to be ushered in by a 'crisis'. The bill amounts to an almost total end of the right to asylum. On the day of its introduction at Westminster the prime minister, Rishi Sunak, made the traditional visit to Dover to declare we must 'stop the boats.' The home secretary, Suella Braverman, spoke of how we must stem the tide of asylum seekers – 'a hundred million' of them as she preposterously and shamelessly claimed.

The government's outrage at trafficking is entirely cynical, its faux concern for the plight of exploited refugees shamelessly contradicted by policies such as 'pushback' and the lack of any sanctioned routes into the country which would enable them to make their legitimate claim for asylum. It is the asylum policy of successive governments that has created the conditions in which trafficking can flourish and tragedies occur. The Home Office requires physical presence in the United Kingdom before an asylum application can be lodged, yet blocks off all 'legal' means of entry to do so. The Nationality and Borders Act went further by criminalising all forms of so-called 'irregular' entry. When twenty-seven asylum seekers drowned in the Channel in November 2021 after their small, flimsy, inflatable dinghy sank, Boris Johnson described himself as appalled and accused France of letting human traffickers 'get away with murder', but the responsibility for these deaths lay with his government. If entry to the United Kingdom for the purpose of claiming asylum was permitted, the traffickers' business model would crumble.

Human trafficking implies compulsion, but refugees are desperate for the safety of a new home. One of those who drowned in the Channel tragedy of November 2021 was making his seventh attempt to cross. He was not being forced to act against his will, nor were the other twenty-six Iraqi Kurds, Afghans, Egyptians, Ethiopians and Somalis who drowned with him. Some of those who attempt the Channel crossing arrange it themselves, clubbing together to buy their own boat, making their own plans. The Home Office knows this, having prosecuted asylum seekers pictured by drones driving small boats in the Channel. What should this be called – self-trafficking? And traffickers themselves are at least meeting a need, albeit on extortionate terms and with callous indifference to human safety. I'd thought of 'pushback' and removal as reverse-trafficking, but in fact it's the thing itself, the forcible removal of people against their will, a form of illegal trade, a modern version of transportation.

The futility as well as the inhumanity of these policies should be obvious even to the Home Office. 'Pushback' and government trafficking will make little difference to the number of refugees attempting to reach the Kent coast. Displaced people have no choice but to move in search of somewhere safe to rebuild their lives. The vaunted deterrent effect of measures introduced by successive home secretaries have signally failed to work. Refugees will continue to risk hazardous crossings of the Mediterranean and the Channel because they are not migrating in order to better their lives so much as to save them.

Consider the example of the Sudanese refugee, Abdul Haroun, and his epic walk through the Channel Tunnel, clinging to metal brackets on the wall as trains sped past, arrested just half a mile short of the tunnel entrance at Folkestone. Haroun had fled from Darfur to escape the persecution of non-Arab Darfuris by the Janjaweed militia, spent a number of years in a camp on the Sudan-Chad border before making his way through Egypt and Libya, crossing the Mediterranean to Italy, eventually reaching Calais where, a few days later, he jumped over a perimeter fence and entered the tunnel, 'early in the morning before the sun came up'. When arrested that evening, he'd walked for almost thirty-one miles. He spoke no English and, on his arrest, uttered a single word, 'Sudan'. Speaking through an interpreter, he told the police that his walk was 'the only solution'. "Even if you had died?" they asked, to which he repeated it was the only solution.

His case for asylum, assisted by the publicity his walk had attracted, was so compelling that it was granted within twenty-four hours. The Crown Prosecution Service (CPS) nevertheless decided to prosecute him under the Malicious Damage Act 1861, legislation from the early days of railway that made it an offence to obstruct engines or carriages. Asylum seekers have legal protection against prosecution under the 1951 Refugee Convention, and Home Office guidelines are that if an alleged offender has already been granted asylum, or it appears likely they will be, then prosecution is probably not in the public interest. In Abdul Haroun's case, with Eurotunnel demanding 'the full force of the law' and the Folkestone MP, Damian Collins, demanding Abdul be sent back to Calais or Sudan, the CPS took him to court. He was held at Elmley prison in Kent for five months before being released on bail and taken into the private care of a local refugee help organisation.

I met him at Canterbury Crown Court where I attended several of his hearings. Abdul doesn't know his exact date of birth, but he is somewhere around forty. He had acquired some basic English during his months in prison and on bail, but even with a translator, I wondered how he understood the crime from the mid-Victorian era he'd been charged with. It was like being in Taylor House again. Points of law were argued back and forth across the courtroom, Abdul standing there – slim, neatly dressed – entirely bypassed by the proceedings. Eventually, after three court appearances, he was sentenced to nine months' jail, the precise equivalent of the time he'd already been held. And so he 'walked free', as they say, but with a criminal conviction that will shadow his right to remain in the United Kingdom and might prejudice his chances of accommodation, employment and education.

The judge made clear that he was only free to leave the court because of the special circumstances of his case – the fact of his having already been granted asylum I assume – and that harsher sentences would be imposed on others who attempted to enter Britain through the Channel Tunnel. She told him that in walking through the tunnel, he had not only endangered his own life but caused 'enormous inconvenience to a large number of people'. It was hardly news to Abdul that in walking through the tunnel he'd risked his own life. He'd been risking it over several months and thousands of miles as he journeyed from the camp at Kari-Yari dam on the Sudan-Chad border and for many years before that. It's difficult to consider the convenience of others when you're desperately trying to save your own life. What measures can possibly deter someone like Abdul? For him, and for millions of other refugees, the only way of saving your life is to risk it.

Attempts to cross the Channel, whether by stowing away in a lorry, braving the heavy traffic of the English Channel in a small boat or by more desperate measures, such as walking through the Channel Tunnel or even swimming, are widely reported. Less often told is the story of the half-life that many refugees are forced into once they have reached the imagined safe haven of the United Kingdom.

Successive home secretaries, both Labour and Conservative, have repeatedly tried and failed to meet their targets to control the number of asylum seekers entering this country. For all its threats and promises, the Home Office is relatively powerless in this matter and will remain so. Its real power lies in creating and intensifying the hostile environment that confronts the asylum seeker once they have reached the United Kingdom, a slower, more insidious form of 'pushback'. Everyone's journey to find sanctuary involves finding their way across a series of borders set up to block their passage. But on arrival in this country, they are faced with yet more borders stretching out nightmarishly in an ever-lengthening line. These can include the walls of a prison or detention centre; Taylor House, a border at which many are pushed back; denial of the right to work, to education, and obstructed access to healthcare. In other words, the denial of entry to a normal human existence. Stuck in a limbo of waiting, the future that the refugee sought remains out of reach. And there are also those informal borders, the daily clamour of the *Mail* and the *Express*, for example. In the lead-up to the EU referendum, both these papers repeatedly carried front-page articles about immigration, all hostile, moulding and focusing that public opinion in which 'every stranger is an enemy' and which has become the most impermeable border of all. Borders and detention centres exist in the mind as well as on the ground.

Inseparable from this is the border of language. The common and endlessly repeated use of terms such as 'swarms', 'marauders', 'illegals' and 'inadmissables' to describe the journeys of the displaced in search of safety and well-being is abusive. Braverman's description of the small boats making their dangerous crossings of the Channel as an 'invasion' is only the most recent example of such inflammatory language. The world of the asylum seeker is one of sacrifice and hardship, which includes separation from family, grief and the loss of love. Our public language has no space for such feelings, not even for compassion. It desensitises and generates indifference, hostility or contempt. The only time that 'swarm' appears in the Book of Exodus is when describing flies. No human being is illegal.

There is nothing abstruse about this concern with the language we use to describe the large-scale movement of displaced people. Language has become a prime means by which we control our borders, a semantic border which persists long after the asylum seeker has entered this country. The Immigration Act 2016, for example, restricted accommodation support to those termed as suffering 'destitution plus'. Plus what? This grotesque and idiotic neologism is a choice example of the dehumanised lexicon that has developed to thicken and patrol the borders behind which the asylum seeker is held. The public language used to describe the tragic exodus of our times has the effect of persuading us not to see what is really being talked about.

Meanwhile, official borders have been thickening all the while. A series of immigration acts which have passed through Parliament, with disturbing ease and very little wider debate, have added literally thousands of changes and hundreds of thousands of pages to the immigration rules, impossible for any applicant to navigate, ready to be turned against the bewildered asylum seeker, forcing them to try and find the punishingly expensive fees needed to have a lawyer argue their case and the Home Office agree to hear it.

The Nationality and Borders and Illegal Migration Acts are now cutting a swathe through these legal obstructions by automatically criminalising the asylum seeker and removing any remaining rights of having their case heard and of appealing against deportation. These successive Acts of Parliament, each tightening the screw of the previous one, codify a deep hostility to asylum seekers and sanction the abuse of their human rights. They are cultural as well as legal measures, giving permission not only to ministers and the courts but to society at large to sustain its hostility to those who make their arduous journeys in search of safe harbour.

When I first came to England in 1969, I spent a term teaching at a comprehensive school in East Finchley, Margaret Thatcher's constituency. Enoch Powell's 'rivers of blood' speech was still echoing, and I remember a group of teachers in the staffroom agreeing that sometime in the future the country would acknowledge that Powell had been right. This has proved not to be so. An opinion poll in 2020 showed ninety-three per cent disagreeing and eighty-four per cent strongly disagreeing with the statement you had to be white to be truly British. But as hostility to those from elsewhere who have already settled has eased, it has intensified to those seeking asylum and has been reinforced by some migrants, as exemplified by several members of the current Conservative government. Once again, the voice of Enoch is being heard, in particular his doctrine of ethno-nationalism, although most now speaking with his voice would disown the association.

This idea of nationality is, of course, a myth. My wanderings through the long history of the East Kent littoral have made clear that migration to this country is at the heart of Britain's island story. Everyone's ancestors have been migrants. I am a migrant myself, though a cushioned one whose right to live in this country troubles no one. I'm Pakeha, not Maori; my accent has modified during the time I've lived here; and most people assume I'm native-born British. But I don't have a British passport, and I certainly fail Norman Tebbit's cricket test of national loyalty. I think of myself as coming from several places: Scotland and Cornwall from where my forebears migrated; New Zealand where I was born and grew up; England where I've spent most of my adult life. Even a story as straightforward as mine has produced plural identities. I feel 'local' in a number of different places. Or do I, any longer? The identification I'd sought with the Kent coastline where I'd come to settle had shifted from place to people, away from locale to those who were being denied the right to cross the border where I live. Can I really feel at home on a tainted seaboard? I'd tramped the coastline into recognition alright, but it wasn't quite the recognition I'd anticipated. I'd seen and learnt too many other things.

When the former New Zealand prime minister, Jacinda Ardern, wearing a hijab, responded to the murder of fifty Muslims at the Al Noor and Linwood mosques in Christchurch in March 2019 by insisting 'we are one, they are us', I thought how impossible it would have been for Theresa May, Viktor Orbán or Donald Trump to make an even remotely similar statement of inclusive nationhood. The victims of this act of terror came from many different places: Syria, Fiji, Pakistan, Gaza, Afghanistan, Mauritius, Kerala, Malaysia, Jordan among others. In embracing these 'strangers', in insisting they were 'us', not 'other', Ardern recognised them for who they were, not what they were. She also spoke of the need for 'a safe and tolerant and inclusive world', one which we don't 'think about in terms of boundaries', challenging the border mentality that currently defines the United Kingdom, Hungary, the United States and many other countries. On the Friday following the killings, the Islamic call to prayer was broadcast on New Zealand's national TV and radio. The BBC would lose its licence for less.

Britain, of all countries, should remember the part played in defending its home ground by those who are now deemed to threaten it. More than a million and a quarter Indian soldiers fought for Britain during the First World War, most of them drawn from the north and north-west, many of them Sikhs, a people long-regarded as inherently warlike and therefore well suited to imperial duty. Perhaps the judge at K's last bail hearing had this in mind when she asked if traditional Sikh weapons had been involved in his conviction for affray. There is an old film of Sikhs wounded in France and hospitalised in Brighton Pavilion being visited by George V. One of these soldiers wrote home: 'No man can return to the Punjab whole. Only the broken-limbed can go back'. Many died and their ashes were scattered on the English Channel. Watching this film, and looking at the British Library's online collection of letters from these soldiers, I thought of K, another Sikh, detained on the cliffs above Dover at the edge of the very same stretch of water.

Primo Levi described how Polish civilians regarded those detained in Auschwitz:

[W]e are the untouchables to the civilians. They think… that as we have been condemned to this life of ours, reduced to our condition, we must be tainted by some mysterious, grave sin. They hear us speak in many different languages, which they do not understand and which sound to them as grotesque as animal noises… They know us as thieves and untrustworthy… and mistaking the effect for the cause, they judge us worthy of our abasement.

Levi, here, captures a condition that others since have analysed. Hannah Arendt, for example, saw the refugee as a figure which, rather than embodying the rights of man, signalled the breakdown of that idea and marked its limits. Giorgio Agamben developed this insight in his idea of 'bare life', the condition of inhabiting 'a space of exception' outside the rights and expectations we normally attribute to human existence. For Z, K and H in the Dover Immigration Removal Centre, Abdul Haroun walking through the Channel Tunnel and for hundreds of thousands of refugees trying to make their way to Europe, the boundary between civil society and bare life has eroded, the only border to do so.

The most glaring and intolerable example of 'bare life' is the plight of child refugees, many of them unaccompanied minors. When Nicholas Winton, organiser of the *Kindertransport* which, in 1939, rescued and brought to Britain hundreds of mainly Jewish Czech children, died in 2015, he was eulogised by David Cameron: 'The world has lost a great man. We must never forget Sir Nicholas Winton's humanity in saving so many children from the Holocaust'. Theresa May, who was Winton's MP, described him at his memorial service as 'an enduring example of the difference that good people can make even in the darkest times', and declared that 'his life will serve as an inspiration for us all... and encourage us to do the right thing'. The contrast between these ringing reminders of how Britain provided sanctuary for refugee children during the Second World War and the indifference of successive British governments to the many hundreds of children who have drowned in the Mediterranean and the Channel, and to all those others cut off from sanctuary by the most unforgiveable of our many borders, is shameless.

Occasionally we are moved to shame. When, in 2015, pictures of Alan Kurdi, the two-year-old Syrian boy whose drowned body was found washed up on the Turkish coastline, were flashed around the world, there was global grief. But the horror and compassion expressed at the tragic death of one little boy extended no further. The deaths of the many, anonymous children who, like Alan, lose their lives when seeking a safe place to live was passed over. In similar manner, after Mo Farah's revelation that he was trafficked as a child, there is no question that his uncertain status in this country will be regularised, as it most certainly should be. But the wider picture – that a very small proportion of child trafficking victims are ever given discretionary leave to remain in this country – continues to be ignored.

The Home Office has repeatedly rejected the applications of unaccompanied minors stranded in France for admission to this country under the family reunification provisions of the Dublin III Regulation, and the Nationality and Borders Act has removed some of these rights altogether, further jeopardising the ability of women and children in particular to reach safety. Unaccompanied children who do make it to this country, whether as asylum seekers or with refugee status, depend on the support of under-resourced local authorities whose actual responsibilities are far from clear. They are not protected by the Children Act 1989 and receive very limited care and supervision. Since 2021 more than two hundred child asylum seekers, many of them victims of trafficking, went missing from hotels where they had been placed in the care of the Home Office and its contractors. Many of them will have fallen into the hands of those who will exploit them. And even those in adequate care become exposed and vulnerable as soon as turning eighteen, the Immigration Act 2016 having withdrawn all care, protection, support and education for children without secure immigration status when they reach that age. Immigration status, not welfare, is always the determining criterion for support.

Most unsupported minors who reach this country as refugees are psychologically and physically damaged, traumatised by their journeys and extremely vulnerable. They are often suffering severe disruption of memory which makes proving their stories very difficult, especially when faced with officials who privilege specificity and consistency above all else. Equally difficult is proving one's age at a period of life when the development of maturity is so uneven. The mere fact of turning eighteen has become yet another border.

New age assessment procedures have been introduced to identify those suspected of pretending to be children, 'age-disputed adults' in the loaded phrasing of the Home Office. These include the X-ray of forearm bones to test the maturity of the skeletal system. Not only is this unethical, indeed abhorrent, but such tests carry at least a two-year margin of error which makes them an unreliable method of establishing, say, the difference between a seventeen- and a nineteen-year-old. Age assessments formerly conducted by Kent County Council social workers are now made 'in house' by Home Office officials, often after telephone interviews and without the young asylum seeker even being seen. Teenagers held in the Manston processing centre reported being pressured by screening officials and security guards to declare they were over eighteen and promised if they did so, they'd be sent to better accommodation. Several succumbed.

A High Court judgement in 2022 ruled that the incorrect assessment of the ages of two unaccompanied male asylum seekers – a sixteen-year-old Kuwaiti who had arrived in the United Kingdom on the back of a lorry and a seventeen-year-old Iranian who had braved an eight-hour Channel crossing in a dinghy – was unlawful as well as mistaken. It is extraordinary how regularly the Home Office, the department of law and order, has been found by international, European and British courts to have broken the law in its treatment of asylum seekers and how such rulings continue to be ignored.

Several Conservative MPs led by the member for South Thanet, Craig Mackinlay, a former UKIP parliamentary candidate, wanted children to be included in the plan to push back the boats. Mackinlay described this as a 'high-octane measure' that would deter young asylum seekers from making their journeys. Did he even pause to consider what he was really saying when using this repellent phrase? As Hannah Arendt wrote: 'Evil comes from a failure to think'.

A month or so later, as if in response to Mackinlay's mindless cruelty, Little Amal walked into Dover after an eight-thousand-kilometre journey from the town of Gaziantep near the Turkish-Syrian border. All the way she'd been searching for her mother who'd gone off one day to find food and never came back. Little Amal is nine years old and almost twelve feet high, a puppet controlled by three puppeteers, one on each arm, a third inside her body. Her personality and emotional life is created through the grace of her movements and the changing expressions of her face – almost but never quite smiling, gentle, composed, curious and determined – by her very large, dark eyes forever searching, and by her huge boots of scuffed red leather.

It was dusk when she walked into Pencester Gardens to be greeted by hundreds of school children waving illuminated stars and by hundreds more adults. She moved slowly through the crowd, her large, expressive hands acknowledging and responding to their welcome, turning to either side, bending down towards those gathered around her. She then led us along Cannon Street towards Market Square, pausing to peer in through the windows of Super Pizza and The Best Kebab and Pizza, hungry perhaps, or else looking for her mother.

The last time I'd been in Market Square, it had been surrounded by a hostile crowd singing 'Rule Britannia', shouting 'close our borders' and 'no illegals', waving banners reading 'Save our Country from Invasion'. The afternoon had ended in violence. Tonight, a large crowd of local people, citizens of Dover who live on the nation's border, had gathered to welcome

Little Amal, a puppet so wondrously animated that she has become a real child evoking compassion for herself and for all those children who, like her, are wandering through Europe. The children of Dover greeted Little Amal as one of their own, as if she were them and they were her, not an unwanted stranger to be pushed back across the Channel. And Amal responded by including them in her plight, recognising through the generous inclusiveness of her movements that some of them might be living their own versions of her story.

I'd gone to Dover that evening expecting the crowd to be small, even wondering if Amal might be met with hostility. Instead, a crowd of several thousand had gathered to offer her their hospitality. Amal's name in Arabic means 'hope'. As I watched the crowd – a man in a high-viz Border Force jacket among them – leave Market Square and make its way up towards the castle where the evening's events would end, the feeling of compassion in Dover that night moved me to hope.

But in the meantime, power lies elsewhere. Borders go up; borders come down, depending on what is at stake. As the free movement of people is resisted and controlled, the free movement of money is welcomed and unregulated. As people try and escape the consequences of invasion, civil war, environmental degradation and other factors for which Western countries bear considerable responsibility, they are grilled about their origins, their motives, their story, their age. Money, on the other hand, conceals its origins and moves with ease across national borders, its story unchallenged. The mendacious rhetoric of scarcity, in which jobs and wages, housing, health, welfare, the nation's prosperity in general is threatened by the comparatively small number of refugees who reach the United Kingdom, has been repeated so often that it has become axiomatic. Meanwhile, money flows in; money flows out; the rich get richer; the poor get poorer; and those with nothing at all get blamed. Offshoring means one thing for the very rich, quite another for the displaced.

A poem of Auden's keeps echoing in my mind, his 'Refugee Blues', written in 1939 at the outbreak of war and the onset of a previous mass movement of displaced people:

> Say this city has ten million souls,
> Some are living in mansions, some are living in holes:
> Yet there's no place for us, my dear, yet there's no place for us.
>
> …

The consul banged the table and said:
"If you've got no passport you're officially dead":
But we are still alive, my dear, but we are still alive.
...
Dreamed I saw a building with a thousand floors,
A thousand windows and a thousand doors;
Not one of them was ours, my dear, not one of them was ours.

14

THE CITADEL

‡

The idea of the concentration camp – the separation and isolation of an unwanted and targeted group – was part of the repertoire of modern colonialism long before it was used by Nazi Germany. The British had employed it in South Africa and the Spanish in Cuba. The Native American reservation and the Aboriginal reserve were further examples of this practice and so too were the leper colonies that proliferated across the colonised world in the later nineteenth and early twentieth century. As Primo Levi understood, the Nazi concentration camps were the extreme end of a continuum rather than a horror without precedent or comparison: 'In every part of the world', he wrote, 'wherever you begin by denying the fundamental liberties of mankind, and equality among people, you move towards the concentration camp system'.

Detention is our sanitised term for the enclosure and concentration of those seeking refuge. A detention centre sounds more neutral than a concentration camp, carrying the sense of a holding station from which those temporarily held will soon be released. But detention centres are literally concentration camps, at the heart of the overwhelmingly security-based response to the twenty-first-century refugee crisis, arresting the movement of the displaced and imprisoning them behind unlimited quantities of barbed wire.

When the philosopher Julian Baggini spent a week as writer-in-residence at South Foreland Lighthouse, I took him up to the Citadel. The barbed wire and surveillance cameras reminded him of a prisoner of war camp, appropriately enough given that Britain's latest war is against those seeking asylum. There was barbed wire everywhere. The Citadel was wrapped in it like some grotesque Christo installation: it topped the high wire fence that reinforced the inner wall of the moat and circled the perimeter of the outer wall for hundreds of yards; it was draped over and around the massive gated entrance and the underside of the bridge that straddles the moat and leads into the Citadel. Leaning over the bridge to examine this barbed wire, the grassy bottom of the moat sloping away towards the edge of the cliff like a slipway, I had a Bosch-like image of detainees being hurled from the bridge, spilling down the moat and dropping into the Channel below.

Barbed wire is a symbol of extreme captivity. It is how we herd together, that is concentrate, the wrong kind of human beings. It indicates a place of coercion, desolation and affliction. Barbed wire – the devil's rope as it is sometimes called – is the stuff behind which the contemporary world holds those it discards. It marks an impassable border, a no man's land.

The United Kingdom has used it in Calais to keep refugees out and in Dover to keep them in. Hundreds of miles of barbed wire border fences have spread across the western Balkans. The southern borders of Hungary and Slovenia are completely fenced off, as are large parts of the border between Turkey and Bulgaria, Bulgaria and Serbia, Greece and Macedonia and Poland and Belarus. Hungary's border wall is especially forbidding, thirteen feet high, stretching over more than five hundred kilometres and topped with the obscenely named 'Nato concertina' – galvanised razor wire, in other words. When the music stops, the displaced are left dangling from the fence.

The Citadel was a place apart, a non-place as David Herd puts it, outside and beyond the norms of civil society that define the nation at whose limit it sat. Those it held were subject to law, though many were held unlawfully and denied the protections that law is designed to provide. Its detainees were deemed not to belong here and were therefore without rights, 'citizens of the world and of nowhere' in Theresa May's disdainful phrase. Indeed, legally speaking, many did not belong anywhere, held in a removal centre but unable to be removed. This was certainly the case with the detainees for whom I'd stood bail. Z's country of origin refused to have him back; K had no address to be returned to; H waited and waited for papers that would restore his lost citizenship, his former national identity, and allow him to leave the country he wished to depart. In terms of jurisprudence a non-place, the Citadel was, however, all too real for those detained there in a Bastille guarding the nation's border.

Herd also discusses how detention centres and the courts of the immigration and asylum tribunal system are 'outside writing'. Every stage of the detainee's experience, from the ban on visitors taking pen and paper into a place of detention to the lack of a full court record of the detainee's bail application and appeal hearings, is characterised by what he describes as a 'recurring limit on inscription'. Official documentation of court proceedings is kept to a bare minimum and records only the voice of the judge, never that of the detainee. The visitor to a detention centre can only carry away information and impressions in their head.

Although on the one hand there is a virtual proscription on paper, on the other there is an overriding official imperative to provide it to prove you are who you are and that you come from where you come from. Accompanying this is the recurring discovery that the piece of paper you have is inadequate, or the wrong piece of paper, that it won't do the business. In Ali Smith's account of her meeting with a former detainee, 'The detainee's tale', the man's ID, his identifying piece of paper, has been ruled insufficient to allow him to enrol at a London college. Yet without this piece of paper, he would sink even deeper into the administrative mire in which he is stuck.

Smith asks to see this insufficient piece of paper, which is also his only form of passport, the sole remaining evidence of who he is. He searches through the two small rucksacks that hold the things he carries, producing at last 'an A4 piece of paper, a photocopy whose ink is creased and flaking, beginning to disintegrate in the folds'. Smith takes the paper and wonders: 'What kind of a life are we living on this earth when a photocopied piece of paper can mean and say more about your life than your life does'. To lose this piece of paper, or to have it taken away from you, means your freedom will also be taken away and you will become lost in indefinite detention, or reduced to bail life with its constant fear of rearrest and further detention, or to bare life in the shadow world of the detainee who goes to ground. Being *sans papier* leaves the asylum seeker without an official identity, stripped of what Frances Stonor Saunders calls a 'verified self'. And as Z found, even when the correct piece of paper – the warrant for his arrest – was retrieved and produced in court, it was turned against him. Detention and the threat of removal is an endless and unwinnable paper chase.

Detention centres are out of time as well as out of place. The United Kingdom is the only country in Europe, including even Hungary, without a statutory time limit on the length of detention. Someone serving a fixed term of imprisonment can count down the days to their release. Those held in indefinite detention can only count up the days as they mount. Time is stripped of its value and the life of the detainee is brought to a halt as their future is walled off. Space and time are the co-ordinates by which we live, that tell us who we are and where we come from and allow us to shape our consciousness of being human. To live out of time in a non-place is to have your life shrivel to mere physical existence, to feel your identity eroding. Time becomes meaningless as Primo Levi described so vividly:

> For living men, the units of time always have a value... but for us, hours, days, months spilled out sluggishly from the future into the past, always too slowly, a valueless and superfluous material, of which we sought to rid ourselves as soon as possible... For us, history had stopped.

Z and K responded differently to prolonged detention, Z with quiet desperation, K with resigned, sometimes bitter patience. But the underlying effect of being caught in a system that seemed to have them trapped whichever way they turned was similar. Dickens, in *Little Dorrit*, calls it 'jail-rot'. Like William Dorrit, held in the Marshalsea until his debts are paid, Z and K had no idea when or how they might be released. Dorrit is eventually freed by that standby of the Victorian novel, an unexpected inheritance, but no such good fortune would come to their aid. Indefinite detention is a form of social death in which your personality and identity crumbles, hope vanishes, existence becomes meaningless.

In 2012, Nana Varveropoulou, herself an immigrant, managed to set up a photography workshop for detainees held at Colnbrook Immigration Removal Centre which lies close to Heathrow, positioned like the Citadel at a ready point of exit. She assembled a record of life inside Colnbrook, some of the photos her own, most of them taken by detainees. As she said: 'How could I ever tell the story of what it's like to be in detention. The only way to communicate the experience was through them'.

Varveropoulou's exhibition of these photos, 'No Man's Land', brought home to me how little I had really understood the life of Z, K and H inside the Citadel. One of the photographs looked down onto a dark tabletop scattered with white dominoes, some of the tiles turned over, blank, others face up, their black dots like staring eyes. Another looked upwards from an outdoor exercise yard at clouds in a blue sky seen through a mesh of high fences topped with barbed wire. The accompanying text by the photographer, M. Noor, remarked: 'There is barbed wire and netting everywhere, even on the sky'. We take an unrestricted view of the sky for granted. Alongside these images of entrapment and isolation were others that gave a more impressionistic feel of the experience of indefinite detention, blurred images that seemed to express the dissolution of selfhood: a kicked-about green football on a dark patch of asphalt for example, seen like the dominoes from directly above, looking like an object in space, lost or forever circling in the same orbit. This had been taken by C Gomez, whose comment read: 'You are constantly surrounded by people and noise, yet you feel utterly alone'.

The detainee whose story Ali Smith recounts describes his time in detention in similar terms: 'There's no privacy… no religious privacy'. I'd observed the effects of prison rot and social death on Z, K and H, but these photographs deepened my sense of the psychological erosion caused by a trapped life of isolation and regimentation, by the combination of institutional noise and existential silence. In the outside world, we can, if we wish, choose to find our company in solitude. The comfort of solitude is beyond reach of the lonely detainee.

On my visits to Z, K and H in the Citadel, we sometimes talked about their past, more often about their prospects for a future, but rarely about their day-to-day life in detention. It seemed pointless, embarrassing even, to ask about an existence that was so bleakly unchanging, in which every day was the same day. Visiting them would remind me of going to see someone who was terminally ill. In both cases the person concerned lacked a future; their life was end-stopped. The difference, of course, is that detention is human and avoidable, death natural and unavoidable. The sequence of feelings I've experienced when visiting detainees or dying friends is however very similar: apprehension and nervousness about making the visit; the edgy intensity of being face to face with someone who has no future; the conflicted feeling of wanting to leave and wanting to stay; the relief and release of departing; the uneasy comfort of resuming one's own untrammelled existence.

Somewhere at the centre of this tangle of feelings was my awkward awareness of trying to bring comfort, even assistance, to the detainee when there was so little I could actually do. Deprived of warm human contact for so long, when offered, it raised the detainee's hopes beyond any likely chance of fulfilment. The result was a complicated sense of guilt and embarrassment: at being free when they weren't; at knowing that to be friendly and sympathetic might mislead them by raising their hopes; at therefore offering false comfort, aware that while humanly speaking this comfort was appreciated, legally and administratively it was powerless to make much difference. Indeed, at Taylor House, if I seemed a mere bleeding heart rather than an honorary probation officer, it could tell against a bail application.

The French philosopher Simone Weil, who died in a TB sanatorium at Ashford, just inland from Dover, in 1943, had a particular word for the condition in which indefinite detention had left Z, K and H – 'affliction' – which she distinguishes from 'suffering' in that it mutilates a whole person's being, causing physical pain, distress of soul and social degradation all at the same time, making a thing out of a human being while they are still alive. The afflicted, she wrote, 'have no words to express what is happening to their them'. Z, K and H existed in a wordless state of incommunicable affliction. But the exhibition 'No Man's Land' gave me a glimpse of the inner condition of the detainee, of their bare life, of an existence from which their rights as a human being had been withdrawn. The photos had expressed more than words could.

In the same week as I was finishing an early draft of this book, the Dover Immigration Removal Centre closed. This came without warning. The staff first heard of losing their jobs one morning on Facebook and local radio. There was no official explanation of the abrupt closure, though one reason seems to have been the shortage of prison staff on the Isle of Sheppey. Prison service employees at the Citadel were given the option of immediately transferring there. Within a couple of days, the three hundred detainees had been dispersed to other detention centres around the country, leaving only one, The Verne on Portland, in the hands of the prison service. Closing the Citadel was another step in the outsourcing and profiteering of the asylum system.

Several months later, I returned to the Citadel, curious to see what was happening to it. The car park was empty, the reception shed locked, the Citadel as forbidding as ever. I felt again the anxiety I'd experienced with each visit, a sense that once I'd crossed the bridge and passed through the small door in the massive gate, I might never come out. The bridge was still girded in barbed wire. As I leant over its side, once more looking down into the moat, a uniformed officer emerged from the small door. He was civil enough, just curious as to why I was there, so I explained that I'd been a visitor at the detention centre and had wondered what was happening to the Citadel. He didn't respond to my question but instead asked me what had happened to the detainees I used to visit. So I mentioned that K, still on long-term bail in

North Shields, had given up and was waiting to be removed. "What a laugh, eh!" he replied. "Comes all that way, costs us all that trouble and money and then pisses off back home." There was scorn in his voice, contempt in his face.

CODA

‡

I'd set out from Dover to Folkestone on the last leg of my water's edge walk, crunching along the pebbles of Shakespeare Beach, Admiralty pier at my back, the Citadel out of sight above me, the massive uplift of Shakespeare Cliff looming in front. At the end of the beach where the cliff lifts off, banking steeply, my way was blocked by mounds of chalk. An avalanche had scooped tonnes of the stuff from the cliff-face and flung it over the rocks and out into the sea. I began clambering.

The fall had eaten many yards into the cliff, leaving a newly whitewashed face, pure and rich. There were deep cracks running up and across it, into which the rains will filter and bring down more. A great jutting piece of chalk, like a flying buttress, had cracked around its base. The concavities produced by the fall had left precarious crags and irregular galleries of crumbling chalk. Shakespeare Cliff is falling down.

In *King Lear*, Edgar describes the scene from the top of the cliff to his blinded father, Gloucester:

How fearful
And dizzy 'tis to cast one's eyes so low!
The crows and choughs that wing the midway air
Show scarce so gross as beetles; half way down
Hangs one that gathers sampire, dreadful trade!
Methinks he seems no bigger than his head.
The fishermen that walk upon the beach
Appear like mice, and yond tall anchoring bark

Diminish'd to her cock, a buoy
Almost too small for sight. The murmuring
Surge,
That on th'unnumber'd idle pebble chafes,
Cannot be heard so high. I'll look no more,
Lest my brain turn, and the deficient sight
Topple down headlong.

Once again I'm hit by vertigo from below, as if I'll topple backwards or even upwards. Edgar and Gloucester's world is strangely mixed with mine, the imagined samphire gatherer no bigger than his feet, and me at the foot of the cliff, a 'cowrin, tim'rous' mouse.

Eventually breaking out of the chalk, I'm back on a pebble and rock beach. Here, directly beneath my feet, is the Channel Tunnel, the largest of the underground workings along this coast, the biggest subterranean world of them all. When catching the train from the Eurotunnel terminal on the edge of Folkestone, you assume you're plunging directly under Dover Strait, but in fact, the tunnel sweeps inland and back almost to Dover before swinging under the A20 and diving into the Channel. I'm walking directly over where Abdul Haroun made his underground journey. I wonder if the trains will in time disturb and undermine the cliff and I stand around for a few minutes, curious to see if they shake the ground in the way that you can sometimes feel the London underground under your feet. But I'm very conscious of the tide hurrying near, and so I press on.

It's not far before I hit another chalk wall. From a distance, it looks like spilt flour, but trying to scramble over and through it is like being caught in pack ice, great globs of chalk all around me, no sign of the level shore, just the sea to one side and the cliffs to the other. I clamber up and down and around for a while, engulfed in chalk, coming to rest on top of a pile of it fifteen or twenty feet high. From here, I can see the low outcrop of Samphire Hoe ahead and I know that, unless I've misjudged the tide, the way beyond to Folkestone will be flat and easy. But it's not the tide that makes me turn back. Between me and the Hoe is yet more chalk, an older, greyer and steeper fall. There is no way around and no way up. The cliffs curl over me like the crest of a nightmarish wave.

Samphire Hoe was created from chalk marl excavated while digging the Channel Tunnel. A low, spreading promontory, a different kind of polder, entirely man-made from this reclaimed spoil, it's another example of the incessant reconfiguring of this shoreline. The bit of coast from where it now spills into the Channel was once part of the old road between Dover and Folkestone before a large cliff fall made it impassable. Some of the cliff was blown up in the 1840s while building the railway between these two towns. And it was here in the 1880s that the first attempt at building a Channel tunnel was made, and a decade later that the first Kent coal mine was sunk. Dug up, blown up, tunnelled, excavated, it has now been extended and reshaped into Samphire Hoe, a chalk meadow mull on which rare plant species, especially orchids, grow and at the back of which samphire and rock sea lavender flourish on the cliff. It is rich in bird life, has walking and cycling trails and a scattering of art installations. Like Betteshanger Country Park created from a coal tip, and Fan Bay and its surrounds reclaimed by the National Trust, Samphire Hoe is another example of the regeneration that is part of the narrative of this coastline.

Defeated by falling chalk, I stand there looking up at the White Cliffs, thinking how, whether standing or fallen, they're a bulwark, a barrier, a buttress, these ugly emphatic plosives spitting out the rejection they imply. So, I take the high road to Folkestone instead, walking atop the stuff that has prevented me from following the line of the shore. The climb up to the top of Shakespeare Cliff is the steepest I've met since setting out from Reculver. There are signs warning of crumbling cliffs and directing me away from the cliff edge. I stand looking at the upwards curl and abrupt drop of the cliff, hesitate and then snake on my belly towards its crest but, a yard or so short of the edge, I freeze. Fearful and dizzy indeed it is.

The fractal geography of this coastline means that its headlands often seem much closer than they are. From the cliff above Samphire Hoe, Folkestone looks as if it's just around the next bend, but it recedes as I follow the gnawed line of the coast. Sun and sky are misting into each other, the coast of France almost impalpable, its outline dissolving like one of Cézanne's late watercolours of Mont Sainte-Victoire, a post-Brexit view of Europe, the continent fading from sight. I think of Coleridge and Lawrence walking their coastlines in times of conflict, unsettled, alienated. Gulls swoop and dive like Spitfires in a dogfight. A kestrel hovers above me like a drone.

The familiar remains of two world wars are scattered along the path: lookout points, gun emplacements, the shell of an old barracks. Approaching Capel-le-Ferne, I come across another sound mirror. Unlike those at Fan Bay, this one stands on top of the cliff, isolated and exposed rather than dug into the side of a hill. It has stood here for nearly a century, cracked and weathered now, haggard and forlorn, graffiti-smeared but also noble, like a *Rapa Nui* (Easter Island) *moai* (monolith) staring out to sea. At Capel-le-Ferne, the path crosses the grounds of the Kent Battle of Britain Museum where a replica Spitfire and Hawker Hunter sit facing the Channel as if ready to re-enact old skirmishes.

There's a photograph from the Second World War of Corporal Sidney Cocks of the 5th Folkestone Home Guard standing alone with a First World War Lee-Enfield Mk1 on these cliffs at Capel-le-Ferne, guarding the nation. From the shape of the cliffs, he must have been patrolling near the sound mirror. Corporal Cocks is shorter than his rifle and, against the great drop of the cliffs, is like a small fledgling perched insecurely at the top of a tall tree. The picture is touching precisely because of the disproportions it captures, one small man facing the immensity and indifference of the human and natural world, a speaking sight that can stand for this whole coastline. The sound mirror is much larger than Corporal Cocks, bigger than those at Fan Bay as well, but it too is dwarfed, by history rather than scale, a stone-age Sky dish out of time but still defiantly, movingly, in place.

It's late summer, and the wild flowers are drying and dying. Butterflies, though, continue to loop and skitter around them. On my way back, just a few yards beyond the sound mirror, I notice, lying flat in the grass beside the path, a bronze sculpture of a book of natural history, open at a page with illustrations of the early spider orchid. It includes an elegantly engraved handwritten description of how this orchid ensures pollination by mimicking the shape of an insect and secreting pheromone-like chemicals to attract other insects. The orchid has two pollinia hanging down at the top of the flower and, as the insect enters, they stick to its back, breaking off as it departs. I marvel at the intricate subtlety of this, quite the most exquisite case of adaptation I've come across anywhere on my walks. The sculpture is one of a series – 'Flora Calcera' – that the artist, Rob Kesseler, has placed along the path between Dover and Folkestone. Inspired by the collections of pressed dried plants he's seen in the herbarium at Kew Gardens, he thinks of his sculptures as pages that have spread like wind-blown pollen along the path.

Back at Shakespeare Beach, I wander along Dover's Western Docks and out onto the dog-leg of Admiralty pier. Marina Lewycka, in her novel *Two Caravans* (2007), describes migrant workers – Bulgarians, Ukrainians,

Serbians, Africans – arranged in separate national or ethnic groups along this pier. There are scores of people fishing here today – Chinese, South Asian, African, similarly grouped – and whoops of excitement when a small silvery-grey fish, less than a foot long, is hooked and reeled in.

At the end of Admiralty pier, I look back along the coastline at the nation's exoskeleton. I can see Folkestone pier to the south and South Foreland Lighthouse to the north, the two joined by those White Cliffs that have for so long figured as the expression and bastion of island nationhood. And I look outwards across the Channel to the Goodwin Sands and beyond, water where once there was land, to those white cliffs on the other side, the matching jigsaw piece of the French coast that has become one of Britain's proliferating borders, even as it turns its back on Europe.

BIBLIOGRAPHY

✦

Abercrombie, Patrick, *East Kent Regional Planning Scheme Final Report* (1928)

Adam Smith, Janet, *John Buchan: A Biography* (London: Rupert Hart-Davis, 1965)

Agamben, Giorgio, *Homo Sacer: Sovereign Power and Bare Life* (Stanford University Press, 1988)

Aldington, Richard, *Death of a Hero* (London: Hogarth Press, 1984)

Aldridge, Alfred Owen, *Man of Reason: The Life of Thomas Paine* (London: The Cresset Press, 1960)

Anderson, Janice & Swinglehurst, Edmund, *The Victorian and Edwardian Seaside* (Country Life Books, 1978)

Arendt, Hannah, *The Origins of Totalitarianism* (New York: Harcourt Brace Jovanovich, 1973)

Ash Bunker: A Brief History 1949–2009, www.thebunker.net

Auden, W.H., *Selected Poems*, ed., Edward Mendelson (London & Boston: Faber & Faber, 1979)

Ballantyne, R.M., *The Lifeboat: A Tale of Our Coast Heroes* (London: James Nisbet, 1864)

Ballantyne, R.M., *The Floating Light of the Goodwin Sands* (London: James Nisbet, 1870)

Betjeman, John, *Collected Poems* (London: John Murray, 1958)

Bills, Mark & Knight, Vivien, *William Powell Frith: Painting the Victorian Age* (New Haven & London: Yale University Press, 2006)

Bloch, Michael, *Ribbentrop* (London: Bantam Press, 1992)

Bradley, A.G., *England's Outpost: The Country of the Kentish Cinque Ports* (London: Robert Scott, 1921)

British Library, online collection of war correspondence, www.bl.uk/world-war-one

Browne, Sir Thomas, *Urne-Buriall* and *The Garden of Cyrus*, ed. John Carter (London: Cambridge University Press, 1958)

Buchan, John, *The Thirty-Nine Steps* (London: Vintage, 2011)

Bulaitis, John, 'George Christopher Solley of King's End Farm, Richborough: Farmer, Entrepeneur, Historian, Fascist', paper delivered at Centre for Kent History and Heritage conference, *Richborough Through the Ages*, Canterbury Christ Church University, 25 June 2016.

Butler, Robert, *Richborough Port* (East Kent Maritime Trust & Ramsgate Maritime Museum, 1999)

Carroll, Lewis, *The Annotated Alice: Alice's Adventures in Wonderland* and *Through the Looking Glass*, ed. Martin Gardner (Penguin Books, 1970)

Catford, Nick, *The Ramsgate Tunnels: Main line Public AirRaid Shelter and Scenic Railway* (Ramsgate: Michaels Bookshop, 2005)

Cobbett, William, *Rural Rides*, ed. Ian Dyck (Penguin Classics, 2001)

Coleridge, Samuel Taylor, *Poetical Works* (Oxford University Press, 1969)

Coleridge, Samuel Taylor, *Collected Letters*, Vol. 1, ed., Earl Leslie Griggs (Oxford: Clarendon Press, 1956)

Collins, Wilkie, *The Moonstone* (Penguin Books, 2012)

Collyer, David G., *Deal and District at War* (Stroud: The History Press, 2009)

Collyer, David G., *East Kent at War* (Stroud: The History Press, 2010)

Corbin, Alain, *The Lure of the Sea: The Discovery of the Seaside in the Western World 1750–1840* (Penguin Books, 1995)

Defoe, Daniel, *The Storm*, ed. Richard Hamblyn (Penguin Classics, 2005)

Dickens, Charles, *Great Expectations* (Penguin Books, 1972)

Dickens, Charles, *Little Dorrit* (London: Dent Everyman, 1969)

Dickens, Charles, 'The Tuggses at Ramsgate', *Sketches by Boz*, ed. Dennis Walder (Penguin Classics, 1995)

Doring, Heike, *From the margins to the centre and back: Trajectories of regeneration in two marginal English coalfields*, unpublished PhD thesis (Cardiff University, 2009)

Eliot, T.S., *Notes Towards the Definition of Culture* (London: Faber & Faber, 1972)

Eliot, T.S., *The Poems of T.S. Eliot*, Volume 1, Collected and Uncollected Poems, eds. Christopher Ricks & Jim McCue (Baltimore: Johns Hopkins University Press, 2015)

Eliot, Valerie, ed., *The Letters of T.S. Eliot*, Volume 1, 1898–1922 (London: Faber & Faber, 1988)

Eliot, Valerie, ed., *The Waste Land: A Facsimile and Transcript of the Original Drafts* (London: Faber & Faber, 1971)

Evans, Nick, *Dreamland Revived: The Story of Margate's Famous Amusement Park* (Bygone Publishing, 2015)

Fleming, Ian, *Chitty-Chitty-Bang-Bang* (London: Pan Books, 1968)

Fleming, Ian, *Goldfinger* (London: Penguin Books, 2006)

Fleming, Ian, *Moonraker* (London: Penguin Books, 2006)

Ford (Hueffer), Ford Madox, *The Cinque Ports: A Historical and Descriptive Record* (Edinburgh: William Blackwood & Sons, 1900)

Former RAF Ash, http://wikimapia.org/4857496/Former-RAF-Ash

Forster, E.M., *A Passage to India* (Penguin Books, 1988)

Gabriel, Mary, *Love and Capital: Karl and Jenny Marx and the Birth of a Revolution* (New York: Back Bay Books, 2011)

Gilmour, David, *Curzon* (London: John Murray, 1994)

Gordon, Lyndell, *Eliot's Early Years* (Oxford University Press, 1977)

Great Britain: Ministry of Defence, *The United Kingdom Defence Programme: The Way Forward* (London: Her Majesty's Stationery Office, 1981)

Gurnah, Abdulrazak, *Scenes from Provincial Life* (BBC Radio 4, 15 January 1988)

Hamilton, James, *Turner: A Life* (London: Sceptre, 1997)

Harkell, G, 'The Migration of Mining Families to the Kent Coalfield Between the Wars', *Oral History*, 6:1 (1978)

Hart, Derek S., *Dangerous Coastline 1939–1945: A Wartime Childhood in Birchington, Kent*, ed. Jennie Burgess (Ramsgate: Michael's Bookshop, 2009)

Hawkes, Jacquetta, *A Land* (London: Collins, 2012)

Herd, David, *All Just* (Manchester: Carcanet Press, 2012)

Herd, David, 'The View from Dover', *Los Angeles Review of Books*, 3 March, 2015

Higgins, Charlotte, *Under Another Sky: Journeys in Roman Britain* (London: Jonathan Cape, 2013)

Hill, Rosemary, *God's Architect: Pugin and the Building of Romantic Britain* (London: Allen Lane, 2007)

Hill, Rosemary, *Pugin and Ramsgate* (Ramsgate: The Pugin Society, 2004)

Hillier, Bevis, *Young Betjeman* (London: John Murray, 1988)

Hillier, Caroline, *The Bulwark Shore: Thanet and the Cinque Ports* (London: Eyre Methuen, 1980)

Hippler, Thomas, *Governing from the Skies: A Global History of Aerial Bombing* (London: Verso, 2017)

Hollingsworth, J.P., *Those Dirty Miners: A History of the Kent Coalfield* (Catrine: Stenlake Publishing, 2011)

Holmes, Richard, *Coleridge: Early Visions* (London: Flamingo, 1999)

Holyoake, Gregory, *Deal: Sad Smuggling Town* (S.B. Publications, 2007)

Holyoake, Gregory, *Deal: All in the Downs* (S.B. Publications, 2008)

Hopkins, Gerald Manley, *A Selection of His Poems and Prose*, ed. W.H. Gardner (Penguin Books, 1964)

Hopkins, Gerald Manley, *Journals and Papers*, ed., Humphry House & Graham Storey (London: Oxford University Press, 1959)

Humphreys, Roy, *Thanet at War 1939–1945* (Stroud: Alan Sutton, 1992)

Immigration Act 2014, www.legislation.gov.uk/ukpga/2014

Immigration Act 2016, www.legislation.gov.uk/ukpga/2016

Ishiguro, Kazuo, *The Remains of the Day* (London: Faber & Faber, 1989)

Jamie, Kathleen, *Findings* (London: Sort of Books, 2005)

Jamie, Kathleen, *Sightlines* (London: Sort of Books, 2012)

Keane, John, *Tom Paine: A Political Life* (London: Bloomsbury, 1995)

Keats, John, *The Complete Poems*, ed. John Barnard (Penguin Classics, 2006)

Kent Underground Research Group, *Kent and East Sussex Underground* (Rainham: Meresborough Books, 1991)

Kinkead-Weekes, Mark, *D.H. Lawrence: Triumph to Exile 1912–1922* (Cambridge University Press, 1996)

Larkin, Philip, *Collected Poems*, ed. Anthony Thwaite (London: The Marvell Press & Faber & Faber, 1990)

Larn, Richard & Bridget, *Shipwrecks of the Goodwin Sands* (Gillingham: Meresborough Books, 1995)

Leach, Derek, *Dover's Caves and Tunnels* (Dover: Riverdale Publications, 2011)

LeGear, R.F., *Margate Caves Cliftonville* (Kent Archaeological Society) www.kentarchaeology.ac

LeGear, R.F., *Underground Thanet: Quarries, Shelters, Tunnels and Caves* (Birchington: Trust for Thanet Archaeology, 2012)

Lellenberg, Jon; Stashower, Daniel & Foley, Charles, eds. *Arthur Conan Doyle: A Life in Letters* (London: Harper Press, 2007)

Levi, Primo, *If This Is a Man* (London: The Bodley Head, 1965)

Lewis, D.S., *Illusions of Grandeur: Mosley, Fascism and British Society, 1931–1981* (Manchester University Press, 1987)

Lewycka, Marina, *Two Caravans* (London: Penguin Books, 2007)

Lycett, Andrew, *Ian Fleming* (London: Phoenix, 1996)

Lycett-Green, Candida, ed., *John Betjeman: Letters* (London: Methuen, 1994)

Mabey, Richard, *Nature Cure* (London: Pimlico, 2006)

Mabey, Richard, 'On the Virtues of Dis-enchantment', in Evans, Gareth & Robson, Di, eds., *Towards Re-Enchantment: Place and its Meanings* (London: Artevents, 2013)

Macdonald, Charlotte, 'The First World War and the Making of Colonial Memory', *Journal of New Zealand Literature*, 33.2 (2015)

Macfarlane, Robert, 'A Counter-Desecration Phrasebook', in Evans, Gareth & Robson, Di, eds., *Towards Re-Enchantment: Place and its Meanings* (London: Artevents, 2013)

Marsh, Patricia Jane, *The Enigma of the Margate Shell Grotto: An Examination of the Theories on its Origins* (Canterbury: Martyrs Field Publications, 2011)

Millar, Jeremy, 'A Friend to All Nations', Louise Neri ed., *Towards a Promised Land: Wendy Ewald* (Steidl: Artangel, 2006)

Miller, David, 'His Heart in My Hand: Stories from and about Joseph Conrad's Sons', *The Conradian*, 35:2 (Autumn 2010)

Moody, Gerald, *The Isle of Thanet: From Prehistory to the Norman Conquest* (Stroud: The History Press, 2008)

Nagra, Daljit, *Look We Have Coming to Dover!* (London: Faber & Faber, 2007)

Neri, Louise, ed., 'Margate', *Towards a Promised Land:* Wendy Ewald (Steidl: Artangel, 2006)

Ogley, Bob, *Kent: A Chronicle of the Century*, Vols. 1 & 2 (Westerham: Froglets Publications, 1996)

Orwell, George, 'The Lion and the Unicorn', in Sonia Orwell & Ian Angus, eds., *Collected Essays, Journalism and Letters*, Vol 2, 1940–1943 (Penguin Books, 1971)

Orwell, George, *The Road to Wigan Pier* (Penguin Books, 1988)

Overall, Sonia, *The Realm of Shells* (London: Harper Perennial, 2007)

Owen, Keith, *Aylesham Through the Years 1927–1987* (Aylesham Heritage Centre, n.d.)

Owen, Wilfrid, *The Collected Poems of Wilfred Owen*, ed. C. Day Lewis (London: Chatto & Windus, 1974)

Parkin, Di, *Sixty Years of Struggle: History of Betteshanger Colliery* (Deal: Betteshanger Social Welfare Scheme, 2007)

Payne, Christiana, *Where the Sea Meets the Land: Artists on the Coast in Nineteenth-Century Britain* (London: Sansom & Co., 2007)

Peverley, John, *Dover's Hidden Fortress: The History and Preservation of the Western Heights Fortifications* (Dover Society, 1996)

Pitt, M., *The World on Our Backs: The Kent Miners and the 1972 Miners' Strike* (London: Lawrence & Wishart, 1979)

Pointon, Marcia, 'The Representation of Time in Painting: A Study of William Dyce's *Pegwell Bay: A Recollection of October 5th, 1858*' (*Art History*, March 1978)

Pointon, Marcia, *William Dyce 1806–1864: A Critical Biography* (Oxford: Clarendon Press, 1979)

Rainey, Lawrence, *Revisiting The Waste Land* (London: Yale University Press, 2005)

Razac, Olivier, *Barbed Wire: A Political History* (London: Profile Books, 2002)

Reed, John, 'The Cross-Channel Guns', *After the Battle* 29 (Battle of Britain International Ltd., 1980)

Saunders, Frances Stonor, 'Stuck on the Flypaper', *London Review of Books* (9 April 2015)

Saunders, Frances Stonor, 'Where on Earth are you?', *London Review of Books* (3 March, 2016)

Seabrook, David, *All the Devils Are Here* (London: Granta, 2003)

Sebald, W.G., *The Rings of Saturn* (London: Harvill Press, 1998)

Self, Will, 'To the Hoo Peninsula, where Marlow and Magwitch met – but no modern folk ever tread', *The New Statesman* (June 2015)

Selwyn, Francis, *Hitler's Englishman: The Crime of Lord Haw-Haw* (Penguin Books, 1987)

Shakespeare, William, *King Lear* (London: Methuen, Arden Edition, 1961)

Shakespeare, William, *Richard II* (New York: Signet Classic, 1963)

Shelley, Mary, *The Last Man* (Lincoln & London: University of Nebraska Press, 1993)

Sillitoe, Alan & Godwin, Fay, *The Saxon Shore Way: From Gravesend to Rye* (London: Hutchinson, 1983)

Smith, Ali, 'The Detainee's Tale', David Herd & Anna Pincus eds., *Refugee Tales* (Manchester: Comma Press, 2016)

Sontag, Susan, *Death Kit* (Penguin Modern Classics, 2013)

Stommel, Henry, *Lost Islands* (Vancouver: University of British Columbia Press, 1984)

Tambimuttu & Marsh, Richard, *T.S. Eliot* (London: Frank Cass, 1948)

The Ramsgate Society, *The Ramsgate Millennium Book* (n.d.)

Treanor, Thomas Stanley, *Heroes of the Goodwin Sands* (London: The Religious Tract Society, 1892)

Twyman, Mick, 'The Mystery of Margate's Shell Temple', *Bygone Kent*, 27:6 (2006)

Ungerson, Clare, *Four Thousand Lives: The Rescue of German Jewish Men to Britain, 1939* (Stroud: The History Press, 2014)

Wedde, Ian & McQueen, Harvey, *The Penguin Book of New Zealand Verse* (Auckland: Penguin N.Z., 1985)

Weil, Simone, 'The Love of God and Affliction', in *Waiting for God* (London: 2009)

Welby, Douglas, *The Kentish Village of Eastry 1800–2000* (The Press on the Lake, 2007)

Wells, H.G., *The Time Machine* (London: Pan Books, 1953)

Wells, H.G., *The War of the Worlds* (Penguin Classics, 2005)

Williams, Peter, *In Black & White* (PWTV, 2020)

Williams, Raymond, *Culture and Society 1780–1950* (Penguin Books, 1963)

Wilmott, Tony, *Richborough and Reculver* (London: English Heritage, 2012)

Wilson, Bee, 'Musical Chairs with Ribbentrop', *London Review of Books* (20 December 2012)

Winder, Robert, *Bloody Foreigners: The Story of Immigration to Britain* (London: Abacus, 2013)

Wood, Christopher, *William Powell Frith: A Painter and His World* (London: Sutton Publishing, 2006)

ACKNOWLEDGEMENTS

‡

Special thanks to those who have helped directly in the making of this book: Lyn Innes for carefully reading the manuscript, for her encouragement and advice and for many years of friendship and wisdom. Murray Edmond and Anna-Marie Taylor for helping me through a sticky patch and showing me how to push on with my book. David Herd for opening up the world of the detainee, for the example of his own writing and for his friendship. Jane Graham-Maw and Elizabeth Briggs for their suggestions. Sonia Overall for sharing her research on the Shell Grotto and for the vision of her novel, *The Realm of Shells*. Clare Ungerson, whose book *Four Thousand Lives* revealed the story of a transit camp for Jewish refugees I knew nothing about. Jo Harman, whose documentary *Safe Haven* showed me another side of the world of the refugee. Pete Keenan and Christine Oliver, who inducted me into the process of bail applications for detainees.

Many thanks to those old friends who showed an interest in this book and whose abiding presence in my life is sustaining: Henry and Renee Bernstein, Sheila Browne, David Ellis, David Flusfeder, the late and much missed Bart Moore Gilbert, Abdulrazak Gurnah, Wade Mansell, Denise deCaires Narain, Susheila Nasta, Vanessa Smith, Sue Swift, Zahid Warley. Thanks also to The Lapwings for our sorties across the seaboard.

Closer to home: my children Cass, Daisy, Jo; my grandchildren Louis, Otto, Molly, Eliza and Max; my sister, Marion, and my brother, Murray, with love.

Most of all, for just about everything really, my partner, Scarlett Thomas.